Online Web Design

The Click and Easy Guide to Creating Great Web Sites

Kristina Ackley and Hilary Benoit

Impact Publications
Manassas Park, VA

Copyright © 2002 by Kristina M. Ackley and Hilary Benoit. All rights reserved. Printed in the United States of America. No part of this book may be used or reproduced in any manner whatsoever without written permission of the publisher: IMPACT PUBLICATIONS, 9104 Manassas Drive, Suite N, Manassas Park, VA 20111, Tel. 703-361-7300 or Fax 703-335-9486.

Liability/Warranty: The authors and publisher have made every attempt to provide the reader with accurate, timely, and useful information. This information is presented for reference purposes only. The authors and publisher make no claims that using this information will guarantee the reader a perfect Web site. The authors and publisher shall not be liable for any loss or damages incurred in the process of following the advice presented in this book. The information expressed herein does not necessarily represent the opinions of the companies for which the authors work.

Library of Congress Cataloguing-in-Publication Data

Ackley, Kristina M., 1969-
    Online web design: the click and easy guide to creating great web sites /
Kristina Ackley and Hilary Benoit
      p. cm.
    Includes bibliographical references and index.
    ISBN 1-57023-166-4
    1. Web sites–Design. I. Benoit, Hilary. II. Title.

TK5105.888.A25 2001
005.2'76dc21   2001039302

Publisher: For information on Impact Publications, including current and forthcoming publications, authors, press kits, online bookstore, and submission requirements, visit our Web site: *www.impactpublications.com*

Publicity/Rights: For information on publicity, author interviews, and subsidiary rights, contact the Media Relations Department: Tel. 703-361-7300, Fax 703-335-9486, or email: *info@impactpublications.com*.

Sales/Distribution: All bookstore sales are handled through Impact's trade distributor: National Book Network, 15200 NBN Way, Blue Ridge Summit, PA 17214, Tel. 1-800-462-6420. All other sales and distribution inquiries should be directed to the publisher: Sales Department, IMPACT PUBLICATIONS, 9104 Manassas Drive, Suite N, Manassas Park, VA 20111-5211, Tel. 703-361-7300, Fax 703-335-9486, or email: *info@impactpublications.com*

Table of Contents

Dedication

To Steve,
your love and support keep me strong
and
in memory of my Grandfather, John Marshall Chalfant,
thank you—the long road led to great rewards
—Kristina

To my wonderful husband, Robert (Ben) Benoit
for all his encouragement and support
and
to my parents, John and Barbara Hooper,
for everything they've done for me
—Hilary

Acknowledgments

As always, so many people helped us through this exercise in a variety of different ways. We would like to thank everyone who offered insight and advice—every bit proved valuable and helpful. We especially recognize the following, who donated time and thought that helped push this book to creation:

> Jane and Phil Ackley
>
> Gray Chapman
>
> Michael Coleman
>
> Brian Graham
>
> Stella Johnson
>
> Ron and Caryl Krannich
>
> Catherine Lockwood
>
> Lori Masanoff
>
> Colette Strawn
>
> Tim Woods
>
> Suraya Zikria

...and the variety of excellent Web sites and books that provided insight, information, and inspiration, identified individually thoughout this book.

About the Authors

As veteran Web surfers, **Kristina Ackley** and **Hilary Benoit** typify power users of the Internet, utilizing the World Wide Web extensively in both their work and personal lives. They can be contacted via email: onlinedesign@topsitesforyou.com.

Kristina works as communications manager for VeriSign, Inc. She also owns and operates Gryphon Communications, a communications and Web site consulting firm located in Herndon, VA.

As a communications specialist for over ten years, and an online user since 1985, she integrates her two interests to effectively meet personal and professional goals. Prior to her current position with VeriSign, Kristina worked in communications management for several small businesses and nonprofit organizations.

An avid editor and writer, Kristina authored *100 Top Internet Job Sites* (March 2000), as well as feature articles and poetry. She speaks and writes for diverse audiences on topics such as career searching and the Internet.

A founding member of Working to Halt Online Abuse (WHOA), Kristina continues with that organization, as well as the National Association of Female Executives. She writes an online help column for *The Tattling Turtle*, published bi-monthly by the Delta Zeta Northern Virginia Alumnae Association, and serves as Webmaster and vice president, Membership, for the group.

Kristina holds a master's degree in Public Relations Management from the University of Maryland at College Park and a bachelor's degree in English, with a German minor and Writing concentration, from Wittenberg University in Springfield, OH. She now resides in Northern Virginia with her family.

Hilary works as director of Quality for W R Systems, Ltd., an information technology and engineering services company headquartered in Fairfax, VA. She has worked in the information technology (IT) field for more than 20 years, starting as a Cobol programmer back in the seventies. Over the past few years, working more and more with Internet applications, Hilary has targeted quality improvement initiatives at the Web application development process and is particularly interested in usability design issues.

Hilary regularly writes papers and gives presentations on software quality-related topics at IT conferences. She is a member of the American Society for Quality (ASQ), the Internet Society, the National Association of Female Executives (NAFE), IEEE Computer Society, the Software Engineering Institute (SEI), Quality Assurance Institute (QAI), and Mensa.

Hilary holds a master of science degree in Computer Information Systems from Boston University, a Post-Graduate diploma of Technical Russian Translation in Astronomy from North London Polytechnic, U.K., and a bachelor's degree in Russian from the University College of Wales, Aberystwyth, U.K. Her professional certifications include: Software Quality Engineer (CSQE) and Quality Manager (CQMgr) with the ASQ; and Software Test Engineer (CSTE) and Quality Analyst (CQA) with the QAI.

Hilary, originally from Ledbury, Herefordshire, in the U.K., now resides in Northern Virginia with her husband.

Preface

Web site development presents challenging career and personal opportunities for many people, and as the Internet continues to grow, so does the need for qualified Web professionals. The range of Web professionals incorporates a variety of technical and creative backgrounds, such as graphic design, editing, marketing, programming, network administration, and much more. Nearly anyone can join the exciting world of online Web design; with a little bit of creativity, you can hone your traditional skills into Web skills.

People build Web sites for many different reasons: to sell something, to provide information, to share personal interests, to entertain, or simply to establish a presence in cyberspace and feel a part of the twenty-first century. Whatever your reason for building a Web site, you'll want to create the best one possible, in the shortest amount of time, and with the least amount of effort. That's what this book is going to show you—lots of helpful tips and tricks to use the resources already available to you on the Internet and beyond. Since the Internet changes so frequently, we will also maintain a list of updated links on our Web site, <u>www.topsitesforyou.com</u>. Be sure to have this book handy when you access the site, since the section for *Online Web Design* links will be password-protected, using text from the book as rotating passwords.

Whether you're working on your first or fiftieth Web site, this book will expand your resources and creativity by providing quickly and easily accessed resources online. You'll learn how to fully utilize the Internet, the hottest resource for Web site development, from dictionaries of coding definitions to free graphics. This book provides you with "added value" insight into areas not normally covered by similar publications, such as methods of testing and evaluating the newly constructed Web site, ways to effectively promote a Web site presence, and how to successfully use the best Web design resources offered on the Internet. It will also help those of you who manage a Web-building process understand the steps involved so you can communicate with your team. Even the most inexperienced Internet user will be able to build a Web site using existing tools in a very short time.

How This Book Helps You Design an Effective Web Site

The chapters in this book are carefully organized to walk you through the online Web design process. You'll start by determining the Web site you wish to design, move into how to design the site, and eventually learn how to promote the site to the world. Following is a brief synopsis of the book's chapters—if you've advanced past a simple understanding of Web site creation, you may be surprised at some of the resources available to bolster that knowledge:

Chapter 1: Getting Started. This chapter provides you with a brief glimpse of the history of the Internet, and what that means for your Web site. Browse through samples of "excellent" sites, both professional and personal, and learn why we think these sites stand out from the rest. Additionally, this chapter takes a look at legal and copyright issues, provides networking resources, and introduces you to some of the types of Web sites you may want to create.

Chapter 2: The Tools You'll Need. Before you can start creating your Web site, you'll need to identify the tools with which you want to work. Learn all about the differences between text and WYSIWYG editors to figure out your favorite type of Web design tool. Uncover the various graphic design programs that will help you create quality graphics. Or, take the **Click and Easy**™ route by using templates to build your Web site—you only supply the content and categories!

Chapter 3: Good Design Basics. Find out what the experts say, and avoid the pitfalls of bad design. Determine the best practices for Web site design, from alignment to color to successful navigation strategies. Also, build your site's foundation, through some careful organization and a well-planned site map.

Chapter 4: Content to Keep 'Em Coming Back. You know the tools and you've learned the basics. How do you create the bulk of the Web site—the content? Figure out the content to add "stickiness" to your site and keep visitors returning. From content acquisition and management to categories for your consideration, you'll really help your site stand out.

Chapter 5: Code Challenges. Sometimes, you can most easily grasp a challenge by working from the ground up. This chapter serves as a great resource for coding HTML, from the variety of tags available to make your Web pages stand out from the rest to showing the pros and cons of different coding styles and techniques. You'll also learn all about linking and placement.

Chapter 6: Your Best Image. Put your best face forward by learning how to successfully create and manipulate quality images. This chapter also expands on maximizing your scanning abilities, sources for free images online, and animating images. Figure out how to use images to enhance your site and not hamper your visitors through slow loading times.

Chapter 7: Special Effects. When do special effects help, and when do they hinder? This chapter explores the variety of special tools available, including audience-interactive tools like hit counters, search engines, and guest books. Figure out how Flash and Java applets can enhance your site, as well as ways to maximize the effect and minimize the irritation of sounds and animation.

Chapter 8: Troubleshooting Your Efforts. Nothing fails to impress more than a site riddled with typographical errors and broken links. Avoid the pitfalls of poor site design through great testing advice, from browser testing to spellchecking. Also, check out the special online testing resources that let you know firsthand whether your site meets expectations.

Chapter 9: Promoting Your Efforts. Once you build your site, you need to tell the world of its existence! Here's how to maximize your efforts with search engines, from optimization to submission of your site. Learn the pros and cons of link exchanges, starting an online newsletter as a promotional tool and advertising through others. This section also covers the importance of domain name registrations.

Site Selection

How did we choose the sites for this book? We have carefully researched these and other sites, and the selected sites clearly met the following criteria:

- **Ease of use**—the site must be easy to navigate and load quickly; spending hours learning how to use a slow-loading site won't help you on a search for Web resources. Graphic appeal was also a consideration for this category; nobody likes an ugly site!

- **Cost**—we've sought out free services and products. However, several exemplary fee-based services have been added to the list.

- **Timeliness and quality of information**—out-of-date and inaccurate information will hinder, rather than help, your Web design process. The reviewed sites stay on top of the trends and issues you need to know.

- **Effectiveness**—quality, not quantity. All of the information in the world won't help you unless the information is effective in satisfying your needs. These sites meet, and often surpass, their basic goals.

The best part of using the Internet is that you can have fun. Keep this in mind while you're searching for the perfect graphic or idea for your Web site. The opportunities are endless, and the Web provides a variety of sites, resources, and entertainment.

1

Getting Started

You're ready to start a new Web site. After observing the frenetic activity of career programmers, corporations, software engineers, Web designers, Web engineers, and e-commerce gurus strut their stuff on the Internet, you prepare to join the ranks of the "Web published." So, where and how do you get started?

Start with the Internet and an open mind. Browse through different sites and note what you like and what you hate. Do pages that use a lot of graphics and animations initially appeal to your eye, then drive you crazy as they take forever to load? Are you drawn to a quick-loading, text-intensive page, but find that the page is all text (maybe even 10 screens' worth?) and very tiring on your eyes? Or, do you find the perfect home page, only to learn that you can't easily figure out how to navigate within the site? By taking a close look at Web sites created by others, you'll gain a good understanding of how to progress with your site.

During the next step, you need to assess what type of site you want to build. Do you want to create a personal or professional site? The varieties are endless for both categories, as this chapter will further explain. Before you start building your site, familiarize yourself with an overview of the rich and intricate history of the Internet, as well as some of the legal issues you may encounter while constructing a site. This chapter will also help you explore some of the top personal and professional sites to help start your exploration; be sure to check www.topsitesforyou.com for monthly updates to these site addresses.

Internet History in Brief

In 1937, H.G. Wells, author of *The Time Machine*, *The Island of Dr. Moreau*, and *The War of the Worlds*, prophesied a world brain, "where knowledge and ideas are received, sorted, and compared. A network not located in one place." He wasn't far off, and the first steps toward the Internet took place in the sixties, when the Department of Defense's Advanced Research Project Agency started a small network known as ARPANET. Dedicated, high-speed lines enabled the network of four computers to transfer data to each other. Over the next few years, corporations like Bell Labs (now Lucent Technologies) and Honeywell, and schools like UCLA and Stanford, worked on the background infrastructure for ARPANET, programming languages like Unix, which are still widespread, and software applications like Telnet.

Throughout the seventies, ARPANET grew from a four-computer network to over 200 computers in the network, gradually moving away from a strong research and military focus to a more commercial presence. In 1974, Queen Elizabeth II sent out the first royal email, and in 1976, Computer Corporation of America developed COMET, the first commercial email software. Also, during this time, the University of North Carolina developed USENET, a worldwide bulletin board system with online message forums that covered a variety of interest groups. The late seventies heralded a broader interest in email use and network communications.

However, as the eighties witnessed the evolution of ARPANET into a commercial presence and an increase in personal computer purchases, greater numbers of people began to connect online. The early eighties heralded TCP/IP (Transmission Control Protocol/Internet Protocol), through which Internet hosts are con-

Unix—*one of the first operating systems to be written in a high-level programming language*

Telnet—*a terminal emulation program that connects a computer to a network server, enabling the user to remotely administer tasks*

nected. In 1984, the National Science Foundation (NSF) added the NSFNET to the communications process, greatly enhancing the speed and capabilities of the network. The world's faith in the privacy and security of the new network was shaken in 1988 as the first Internet "worm" temporarily disabled approximately 6,000 of the 60,000 network hosts. After the assault, CERT (Computer Emergency Response Team) was established to address future attacks. Shortly thereafter, ARPANET gracefully stepped back, leaving the world with a network of networks known as the Internet. In March 1989, CERN (Conseil Européen pour la Recherche Nucleaire) proposed a new communications tool for members of the widely spread organization. The result was the World Wide Web, a subset of the Internet, a hyperlinked collection of individual pages that are accessed via Internet protocols (URLs) and interpreted through coding languages, such as HTML. Web browsers, such as Netscape and Internet Explorer, translate the language and the links so users can easily navigate between different pages.

During the nineties, the Internet exploded in size and usage, starting with the development of Mosaic, the first graphics-based Web browser in 1993. During the same year, domain names, which enable numeric IP addresses to be translated into alphanumeric names for ease of use in connecting to Web pages, grew at a registration rate of 300 per month. In 1995, the NSF turned the management process over to a consortium of commercial Internet Service Providers (ISPs), and the Internet was developed into its current form.

During the second half of the decade, the World Wide Web became the major function of the Internet. New programming languages like Java and animation programs like Shockwave appeared, increasing the demand for programmers and Web designers. In 1995, the Vatican entered the fray through the creation of its Web site, www.vatican.va, which joined over 120,000 existing domains. Internet usage increased as ISPs offered customers cheaper, faster ways to connect online. In conjunction with

Web browser—*a software application, like Netscape or Internet Explorer, through which you can view a Web site*

improved service, more individuals and companies proceeded with Web page creation, and by 1998, over 300 million Web pages existed, with an expansion of 1.5 million per day. In 1999, Network Solutions announced the 5 millionth domain name, and the Web became an extremely competitive marketplace for products and ideas.

The growth of the Internet's size and outreach has definitely affected the design of Web sites—from content creation, to graphic design, to extended programming languages. Savvy business people stay on top of the Web curve, creating pages that set higher standards for speed, usability, and graphic design. Embrace the expansion of the Internet and the creative growth that follows. Take advantage of this changing landscape and keep informed about top issues concerning legislation, applications, and general development of the Internet.

Internet Legal Issues

Let's not beat around the bush—you've heard the phrase "imitation is the sincerest form of flattery," and you've seen that many Web sites bear a striking resemblance to each other. However, before you step forward and "borrow" a graphic, block of text, specialized coding processes, or site design, make sure you know the laws that could apply. Although the Internet has thus far avoided specialized regulation by the government, laws related to copyright, trademark, freedom of speech, and other issues apply online as well as offline.

One of the most frequent and easily achieved transgressions against intellectual property involves illegally reproducing original works. Both aspiring and established Web designers search the Web for interesting text and graphics that can enhance their sites; however, they frequently fail to gain permission to reproduce the text or graphics. In many cases, gaining permission from the owner of the original material is a simple process and can be achieved through emails. The results of requesting permission vary—from a simple "okay" via email, to a mutual gain scenario, to outright denial. Some examples of mutual gain scenarios include identification of the owner of the original work, adding a link to the property owner's Web site, etc. For more information concerning copyright laws, be sure to check out the following Web sites:

Library of Congress
www.loc.gov/copyright/

The comprehensive resource for copyright information, the Library of Congress provides a variety of information through this site. Browse through general information about copyrights, frequently asked questions (FAQs), and recent news and legislation. You can also search existing copyrights or register a copyright online.

The Copyright Website
www.benedict.com

This site offers a variety of useful information pertaining to copyrights, providing an easily understood introduction to copyright issues, including copyright registration, how to create a copyright notice, and fair-use policies. The Copyright Website also offers a variety of visual, audio, and digital case studies.

Another frequently encountered intellectual property problem involves trademark infringement. Although the types of illegal activity in this area vary, one of the topics widely covered by the media involves domain name registrations. If a company or individual that owns a trademark attempts to register a domain name to strengthen its trademark online and discovers that

ICANN (Internet Corporation for Assigned Names and Numbers)— *the organization responsible for certain issues regarding IP address space, domain name systems, and root server systems*

someone else has already registered a domain name that resembles the trademark, a domain name dispute will most likely occur. Fortunately, in January 2000, ICANN developed the Uniform Dispute Resolution Policy to assist with conflicts in this area. However, try to avoid getting involved in one of these disputes; develop a basic understanding about trademark law, and do your best to ensure you do not violate the law.

U.S. Patent and Trademark Office
www.uspto.gov
As the comprehensive resource for U.S. trademark information, the USPTO supplies a well-stocked Web site that covers everything from searching trademarks to filing trademarks online. Especially notable, the site's Trademark Electronic Business Center walks users through the complete trademark application process.

Trademark.com
www.trademark.com
Trademark.com offers a comprehensive, fee-based solution for trademark-related searches, such as new marks and names, domain names, litigation issues, and competitive intelligence. The site provides a variety of other information, including a great selection of relevant links to international resources for trademark information.

Unfortunately, intellectual property law represents only a portion of the legal issues involving the Internet. Make sure that you carefully read the service agreement you enter with the company that hosts your Web page; in many cases, strict guidelines advise that your Web site is forfeit if you break any laws. To avoid legal ramifications, many Web hosts may close down your site if illegal operations are suspected. It is important to recognize these issues not only to ensure that you do not break the laws, but also to recognize when others have committed a crime against intellectual property that you have created. For more information about additional issues, consult your Web hosting company, or any of the following links:

NOLO Self-Help Law
www.nolo.com
One of the most comprehensive online sources for legal information, NOLO answers a variety of Internet law questions beyond intellectual property, including marketing/advertising, cookies, SPAM, privacy, and much more.

Bitlaw
www.bitlaw.com
In addition to providing great information about intellectual property, Bitlaw explores other Internet law issues, such as service provider liability, Web page linking, and defamation.

So You Think You Want a Web Site— Triumphs and Tribulations

In the perfect world, if you built a Web site, hordes of visitors would magically appear! You'd have no problems or difficulties, and the entire process would be pain-free. However, as we all know, nothing's perfect and you should be prepared for potential pitfalls of Web site development.

Later chapters will go into more detail on good design versus bad design, but be aware of some pretty basic issues that really upset Web site visitors. For example, be careful who hosts your Web site. Hosting sites that inflict irritating pop-up ads on your Web site visitors are guaranteed to take the shine off your site. Avoid typos, spelling mistakes, and poor grammar, since such errors send a message that perhaps your entire Web site is unprofessional. If you feel you need help in this area, you can find plenty of software, as well as Web sites, that can check your site for errors. Make sure you avoid the classic mistakes of Web site design, such as those mentioned in Chapter 3, Good Design Basics, or you could find yourself a target of self-appointed critics who delight in pointing out what they consider to be bad Web sites and then advertising it on the Internet. Do you have what it takes to put up with criticism? If so, put your best foot forward and consider some of the rewarding aspects of creating a Web site.

Just as bad sites can end up on a list on someone else's Web site, good Web sites can also find themselves on the receiving end of surprise accolades. Imagine your joy at discovering that your Web site is on a list of best sites or great designs, or learning that other Web sites want to link to your Web site! If you're building a Web site at work, you may be lucky enough to get a bonus for the site launch. Remember, you don't have to be a programmer or a genius to build yourself a terrific Web site!

The following sites give you an idea of some of the awards that are given to Web sites. Take a look at some of the winners and pick up some hints and tips on good Web design:

World Best Websites
www.worldbestwebsites.com
World Best Websites, well-designed itself, contains a list of awards and judging criteria for its contest, which it describes as being "very tough & a challenge to even the best designers." The site gives instructions on how to enter for an award, lists the judges, and even gives examples of the judges' own work. You can even apply to become a judge!

The Web 100
www.web100.com
User ratings determine the 100 best sites on the Web, listed on this intriguing site. Programs work in the background to compile your ratings and provide hourly updates that rank the sites from 1 to 100. Guidelines and rules for rating Web sites are posted on the site.

Web Sites that Suck
www.websitesthatsuck.com

Learn how to develop great Web sites by studying examples of bad design. The "Daily Sucker" showcases a new Web site each day as an example of poor design. This well-designed site has a wealth of excellent information—articles, chapters from books, and links—on a wide range of Web design topics.

Personal Web Sites

People build Web sites for many different reasons: to sell something, to provide information, to share personal interests and information, or to entertain. Some seek to simply establish a presence in cyberspace and to feel a part of the twenty-first century.

Whatever your reason for wanting to build your own Web site, you'll want to create the best one possible, in the shortest amount of time, and with the least amount of effort. So let's start by taking a look at some of the different types of personal Web sites that you can create. The great thing to remember is that with all the online resources available, it's perfectly possible for anyone to have a well-designed Web site with a minimum of effort.

A personal Web site is a great place to start if you're attempting a Web site for the first time. Take this opportunity to tell the world something about yourself. Enjoy yourself—you're creating a unique, interactive way to highlight yourself or some area of your personal interests. If you cannot figure out what to put on your personal Web site, consider these ideas: photos, family, music, crafts, hobbies, favorite volunteer activities, recipes, movies, books, and pets. Keep in mind that your Web site creates, in effect, a calling card—this is what visitors will see of you, so make a good impression. Remember to check your spelling and grammar!

> **Cyberspace**—*the electronic medium of worldwide computer networks, in which online communication takes place*

Keep your Web site updated, and provide dynamic, interesting resources to visitors—something to keep them coming back. Many Web sites on the World Wide Web don't seem to have organization or focus and, like graffiti, just clutter up cyberspace. Others remain the same five years after their creation, with no new content or technology. Try to make your personal Web site a valuable contribution to the World Wide Web; your visitors will appreciate it!

About.com's Personal Web Pages
personalweb.about.com

This is a great Web portal for information on personal Web sites. You can take a look at other people's personal Web sites, learn how to make money with yours, read valuable tips and tools, and download clipart and other great accessories. For those of you with big ambitions, a special section advises how to promote your site and how to price your Web work.

Online Success for Internet Business
www.webmastercourse.com

This site is easy to navigate and well laid out. The "101 Web site ideas" section is particularly useful if you're trying to think of ideas for your own personal Web site. You can also take an online design course (for a nominal fee) from home and subscribe to the *Web Design Weekly* online newsletter.

Family Communications One of the most popular types of personal Web sites features a topic that you know very well—your family. You can start by describing yourself and members of your family—giving such details as names, birthdays, and hobbies, and including a nice photo. Describe family outings or events that may be of interest to relatives or friends, and maintain a photo gallery. Distant relatives who don't see you often will be able to access your family Web site to see what you've been up to. Taken a step further, you could invite family members to email you information and photos of their activities for the Web site and in this way, you can maintain a communal, internationally accessible family Web site.

Alternatively, a family genealogy Web site can reflect your family history over generations and provide a useful and entertaining source of information for anyone with your family name. Whether you're an amateur or a professional genealogist, you'll find a boon to your research via the wealth of worldwide information available to you on the Internet. Records, databases, and other reference materials are just a mouse-click away and you don't even have to leave your chair. You can download special software that interfaces with hundreds of CDs of genealogical records and references available for a modest fee.

Your personal family history Web site can add to existing traditional genealogical resources for your family and make a real contribution to others who may be researching your family. Add some interesting text, links, and photos of your family as far back as you're able to research. If you add contact information such as your email address, you may make contact with long-lost members of your extended family and enjoy a very rewarding hobby.

MyFamily.com
www.MyFamily.com
Create a free Web site for your family name in approximately three minutes with this site. After a simple site set-up process that involves entering data into a form, you click on the button called "CreateSite."

Genealogy.com
www.genealogy.com
This well-organized site lists tools and resources to help you trace your family history. You can search for a name and view a list of records, databases, repositories, and other resources connected with that name.

Cyndi's List

www.cyndislist.com

This is an excellent portal site with more than 97,700 links and more than 150 categories that encompass a wide range of ethnic and religious roots. You can search on ports of entry, passenger lists, schools, occupations, and many more search criteria in addition to the more usual searches on names and countries. Five different indexes allow you to access information quickly and effectively.

Highlighting Your Hobbies Many people feature their hobbies online—from poetry to opinion pieces, pets to posies, or sewing to music. This type of site enables you to feature samples of how you pursue your hobby. For example, if you like needlepoint, your site could include patterns you have designed, scanned images of successful projects you have completed, and even a video demonstrating your artistry in action!

Not only can you use the Web as your own personal showcase, you can also provide useful resources for others who pursue the same interests. A set of relevant links would offer a way for other people to access information about techniques, vendors, and displays of others' work. Another way to use links to expand your site's reach is to join or start a Web ring; not only will you drive traffic to your site, you'll also help promote other sites that contain similar topics. You could also host a discussion area that enables people with similar interests to share their comments and questions. The potential for hobby sites is limitless. Take a moment to browse through the following sites for more ideas and opportunities:

Yahoo! GeoCities

www.geocities.com

Are you looking for sites that feature sewing, candle-making, horseback riding, or soccer? GeoCities provides an easily searchable directory of the sites it hosts, as well as many other options, should you select this service to host your site.

Tripod
www.tripod.com

Tripod offers another great resource for hobby sites. Not only can you search through others' Web sites; you can also take advantage of the great resources available on this site to create and host your own Web site.

Yahoo! WebRing
www.webring.org

Not only does Yahoo! WebRing show you how to link your site to a related Web ring, it will also help you locate other sites with topics similar to yours. If the perfect Web ring doesn't exist, have no fear—it's easy enough to create a new one!

Professional Web Sites

Professional Web sites can serve as excellent baselines for Web success. Frequently, a team of talented individuals works together to create the site, combining expertise in graphic design, content creation, network management, and advanced programming skills. Web-related jobs can result in lucrative salaries and infinite professional rewards. The scope of these Web sites can run from a resume Web site designed to highlight a person's skills with technology, to large corporate Web sites, to Web portals like Yahoo or Excite.

Perhaps you're leading a team, logging some hours as an outside consultant, or designing a new site with yourself as the sole team member. Regardless of the number of people with whom you work, your budget, or your deadline, building a professional Web site requires a strong attention to detail and quality. A well-built personal Web site will include these factors, but a professional Web site demands them, and not meeting these needs can result in very negative consequences, including lost income or employment.

Web Resumes So, you want to show potential employers your prowess in Web design by creating a quality Web resume? By creating a Web resume, you offer employers a quick and easy interactive method for accessing sites, graphics, flash videos, and a variety of other projects on which you have worked.

Make sure you know how to create a bulletproof Web page before attempting to create a Web resume. Keep in mind that no matter how well you design Web pages, every mistake you make in a resume will be magnified tenfold to a potential employer. Refer to Chapter 8, Troubleshooting Your Efforts, for excellent resources and advice to help ensure that your Web resume portrays the most effective, professional version of "you" possible. Also, remember that many organizations print out and distribute resumes to hiring managers; your brilliant, perfectly designed resume may end up on a piece of paper. With that in mind, make sure that if you provide only a Web resume, you configure it to successfully print, as well.

Try to follow the same formatting guidelines as for regular resumes. Yes, you want to demonstrate your Web design prowess, but don't confuse your end result with having fun. You should provide a resume that employers can easily navigate, that highlights your accomplishments, and that clearly demonstrates to potential employers what you will contribute to their organization. Before venturing into the dynamics of electronic resumes, take some time to update your traditional, paper resume. It will serve as the basis from which you develop other resumes, and **you should always provide a classic, well-organized, nicely designed paper resume during an interview.** Several books provide more in-depth coverage of resume development, and can be located through **www.impactpublications.com**, your local library, or a bookstore:

- *Dynamite Resumes*, by Ron and Caryl Krannich

- *The Savvy Resume Writer*, by Ron and Caryl Krannich

- *Resumes for Dummies*, by Joyce Lain Kennedy

- *Haldane's Best Resumes for Professionals*, by Bernard Haldane Associates

It's up to you whether to include a photo of yourself, but it's safest to avoid doing so. At least 50 percent of employers will dislike some aspect of your photo, and you. The photo provides an employer with things to pick apart—your hairstyle, eyes, color, smile, or attire. Why provide them with that opportunity? Stick with the standards and play it safe. Several online sites focus specifically on resumes tips and tricks, which will help you produce a high quality, well written resume:

4resumes
www.4resumes.com
Filled with an excellent assortment of links to helpful hints and interactive resources, this site will prove extremely helpful toward your resume and cover letter preparation. In addition to listing some of the major resume boards and cover letter and resume writing tips, 4resumes provides links to certified career coaches and reputable guides.

Resumania
www.resumania.com
This entertaining site provides a lighter look at the world of resume writing. Through examples of blunders and mishaps from real-life resumes, this site should help you improve your resume writing skills and avoid making careless errors.

Company Web Sites Congratulations. You're in the "big league" of Web site design—creating company Web sites. Whether you work for a small nonprofit or a large corporation like Microsoft, your task remains the same: to create a dynamic, interesting site to which the masses will flock. A company Web site must set and attain the highest standards—if the links don't work, graphics look bland or poorly designed, or proofreading has not played a major part of the Quality Assurance process, you're in serious trouble. Not only does this site represent your skills and abilities in the Web world, it also portrays the company it represents.

The Web presents limitless opportunities for those who work with professional Web sites. Take a look at some of the major job sites, like Monster.com, Headhunter.net, and Careerbuilder.com, to learn about the skills required for different Web-related positions. You can also try browsing through the Web sites of Web design companies to review the competencies they fulfill. If you lead a team of Web professionals, this research helps you gain a better understanding of the different team roles necessary to pull off a successful site. Of course, salary remains a driving force behind much of the work in the computer sector. Below is a table giving national averages for Web-related positions:

National Salary Data

Title	Low	Median	High
Content Specialist, Sr.	$43,872	$43,872	$61,400
Creative Director—Web	$103,045	$111,200	$124,195
Editor—Web	$43,807	$45,986	$53,158
Producer—Web	$53,267	$61,207	$72,275
Product Manager—Web	$64,961	$74,284	$85,691
Web Designer	$42,283	$52,276	$59,389
Webmaster	$46,849	$54,030	$66,820
Web Software Developer	$49,020	$58,017	$69,043

Reprinted with permission from Salary.com, 05/01

Company Web sites can benefit even more strongly from this book than other Web sites described in this chapter. Browse through some of the larger corporate sites, such as Microsoft.com, Dell.com, and Amazon.com to understand the "competition" more thoroughly—what drives them, standards for excellence, etc. For more information about sites with high ratings, visit:

The Webby Awards
www.webbyawards.com

The Webby Awards, presented by the International Academy of Digital Arts and Sciences and hailed as the "Oscars of the Internet" by worldwide media, identifies the best of the best in Web sites. Categories extend to nearly every type of site, but highlights of professional sites include: Government & Law, Activism, Commerce, Community, Education, Finance, Health, Science, and more.

WebAwards
www.webaward.org

This site, founded by the Web Marketing Association (WMA), seeks to set a high standard for Internet marketing and corporate Web development on the World Wide Web. Award winners included a variety of businesses, including manufacturing, travel, automobile, health care, and many more.

Web Portals In its simplest form, a Web portal displays a collection of links to other Web sites. Usually, the links relate to the same topic. A good Web portal is worth its weight in gold since it saves you a great deal of time. Like reading a magazine that focuses on a specific topic, a good Web portal offers a combination of links, articles, and graphics—constantly updated, easy on the eye, and simple to navigate. Well-known Web portals that also act as popular search engines include Yahoo, Excite, Alta Vista, and HotBot.

GreatStartPage.com—*"Where the Web Begins"*
www.GreatStartPage.com

This site features a customizable start page with links to search engines and other Web sites. A nice feature is that you can add a list of your own links.

Everdene L. L. C.
www.everdene.com

Advertised as "the Portal to the Portals," this well-designed site lists a wealth of popular Web portals and search engines. Other links are listed in categories such as Current Headlines, News, and Weather; Quick Shopping Links; Top Software Download Sites; General Reference; and many more.

Networking Resources

Networking resources abound on the Net, giving you panels of experts for advice on creating a great Web site. Perhaps the most easily recognized are listservs and newsgroups, both of which use email to communicate between participants, and chat rooms, which enable users to conduct real-time conversations about the topics which interest them. In some cases, you can even request reviews from other qualified individuals—but be careful, because the result may provide more responses than you can handle! Positives and negatives exist for these online networking tools, but all three provide useful resources to the active job seeker.

Listservs Perhaps the easiest and most rewarding networking resources, listservs (also known as mailing lists, email lists, or list services) are free and very passive, once you have subscribed to them for the first time. Subscription methods vary, but make sure you read and archive the basic guidelines provided when you sign up for the list, and please do not post marketing and self-promoting messages to the group! In most cases, this can result in your immediate dismissal from the list.

Listservs provide a lively, interesting way to network online. Where else can you connect with hundreds of professionals with similar interests in a nearly instantaneous manner on a forum where everyone can express their opinions and share their experiences? Not only can you gain valuable information about new Web trends, browser compatibility issues, and freebies; you can also bounce off new design and coding ideas among "old pros." Although they're invaluable, listservs do contain some minor pitfalls: Users can digress from the main focus of the list, sometimes generating scores of bickering messages, and subscribers may be overwhelmed by the hundreds of email messages posted to the groups each week. Using a "digest" subscription method condenses the list messages into a single message, reducing the number of emails sent to you. You may be able to determine the frequency with which you receive these messages. Following are some excellent directories and services for researching which listserv to join:

Liszt

www.liszt.com

This dynamic, easy-to-use site provides over 90,095 mailing lists from which you can pick and choose! Liszt recently merged its resources with Topica, a free listserv hosting service. Users can search Liszt by keyword, or can browse through a variety of categories. Liszt also provides a "junk filtering" option, through which users can select to filter irrelevant or private lists from their search, which can produce a maximum of 150 results per search.

Yahoo Groups

groups.yahoo.com

Yahoo Groups also provides an easy, friendly format that users can search for free listservs to which they may subscribe. However, Yahoo Groups takes the search process one step further and enables browsers to search more than 50 million messages, in addition to its 300,000 groups! The Web design category alone produced over 600 results. Additional resources include interactive group calendars, file sharing, and a chat and talk area.

Newsgroups Newsgroups function similarly to listservs by providing networking via email; however, users must actively select the newsgroups in which they are interested each time they seek to access the messages. Newsgroups are rarely moderated, which makes them more susceptible to spam, junk, and irritating pranksters. Information seekers will also encounter an overwhelming number of messages posted to these lists, so please practice some caution in believing everything you read! These sources can provide a variety of interesting and useful information, and should not be discounted based on the freedom of speech that is actively practiced therein.

Your access to newsgroups is determined by your ISP, which subscribes to a selected group of listservs. Don't worry about the resources being too few; many ISPs provide access to so many groups that you can easily be overwhelmed! To start using newsgroups, you'll experience the least confusion by asking your ISP for instructions about connecting to their newsgroup server. Once you have connected to your ISP's service, you can search through a wide assortment of useful and irreverent groups—the key is to keep an open mind while you search for the end result.

If you are frustrated with newsgroups, several sites provide concise, easy-to-understand directions, as well as lists of newsgroups to which you may wish to subscribe:

Google Groups
groups.google.com
Google Groups offers users a quick and easy way to search through the online forums that make up USENET, the

largest repository of newsgroups. Search by a specific keyword to find great networking resources for graphics and coding, as well as a variety of other resources. You can also search through newsgroup messages recorded since 1995.

Liszt's Usenet Newsgroups Directory
www.liszt.com/news

Liszt's newsgroups directory lets you navigate through the variety of available newsgroups. Searching for the word "Web" returned over 140 newsgroup results! The site also contains some great information about using newsgroups.

Professional Organizations Professional organizations, in which people participate based on their profession, industry, or a variety of other interests, span a wide spectrum of associations and membership organizations. Some of these organizations provide membership based on the individual's interests or career; others are based on the organization. Members can participate in networking activities, receive publications related to the industry, opt for special discounts on insurance and other programs, participate in award and recognition programs, study to achieve certifications in their field, and receive many additional bene-fits. For a good head start, visit:

American Society of Association Executives
www.asaenet.org

With over 23,000 members worldwide, the American Society of Association Executives (ASAE) provides a great contact list for associations with specific focuses. Try searching for "graphic design," "Web," or other keywords to determine which associations support your interests.

The Marketing Resource Center

www.marketingsource.com

This Web site provides a "Directory of Associations," which is a searchable database of over 5,000 business-related associations around the world. You can either enter a keyword and search the database, or use the provided drop-down list of association categories to view a list of names and addresses of industry-specific associations.

2

The Tools You'll Need—Page Creators and Image Editors

Now that you know what type of Web site you want to create, you need to determine how you'll create it. In order to create a dynamic, high quality Web site, several types of programs improve your design quality and production speed: one to help you to design and/or "code" the pages, and another to create and/or edit the images for those pages. Depending on your needs, you might also consider programs that let you work with animation, sounds, or fulfill a variety of other tasks; these specialized programs will be briefly featured in later chapters.

Whether you are working on your first or fiftieth Web site, for business or leisure, you can approach the process through many different avenues. Some people prefer to create pages from scratch, using HTML coding techniques for everything from bolding to table design and beyond. Those with a background in graphic design, on the other hand, might approach Web design through software with a graphical interface. Novices to both computer programming and graphic design could select pre-formatted templates...or hire an outside designer!

Fortunately, the Internet supplies a vast amount of resources for these tools and more; great price-comparison resources and online stores eliminate the need to pay full retail for software. Make sure that you purchase your software from a reputable point of sale, or you may run into problems with dishonest sellers or software pirates. Consumer safeguards continue to improve and increase in numbers; however, you need to familiarize yourself with the options available and the risks involved. Look for symbols like the TRUSTe mark for privacy and VeriSign's Secure Site seal to ensure that you are protected and that your sales point cares about ensuring the safety of itself and its customers. As you look through the programs listed in this chapter, keep in mind the following resources for software purchases:

ebay
www.ebay.com

An excellent resource for new and used software, this online auction site gives you a vast assortment of sellers and products. But, buyer beware—although ebay attempts to ensure that its sellers do not distribute illegal copies, a few inevitably get sold. Check out prospective sellers (both individuals and companies) via the clever ratings system, and do your best to buy from reputable sellers.

Computershopper.com
www.computershopper.com

This innovative site lets people compare, evaluate, and buy software and other computer products through a variety of vendors—from leading sources like Egghead.com to small, independent e-sellers. The site provides links to reviews on different products, offering you a great way to determine the bleeding edge solutions you won't want to miss.

Text Editors—Coding from the Hip

Perhaps you're talented enough to understand the ins and outs of directly coding HTML, SHTML, XML, JavaScript, or any of the other coding languages used "behind the scenes" to make Web sites work. At the time of this book's creation, HTML still represents the standard for Web site creation; however, other dynamic languages, such as XML and SHTML, are moving forward into prominent positions in the Web development arena. Popular text editors include HomeSite, Microsoft Notepad, VI (Unix), Emacs (Unix), and BBEdit (Mac), though others also present good options for those who prefer this type of editor.

In Chapter 5, Code Challenges, we present more information about the specifics behind coding with HTML; however, if you're undecided about whether to work directly with code or use a graphical program, take a few minutes to consider the benefits of working with a text editor. First, a text editor enables you to completely control your Web site creation. Since you're working directly with code, **you** determine how the text runs and exactly how different elements are coded. With graphical Web creation software, the program codes the back end for elements, such as tables and forms, leaving you with less initial programming control. Directly creating the code may save you from encountering discrepancies between browser standards, as long as you know those standards. (You'll find out more about browser standards in later chapters or can visit www.w3c.org, the World Wide Web Consortium, for more information.) In addition, a text editor will usually result in fewer computer-related errors; however, you may encounter more user errors if you do not have good coding skills. And finally, most text editors cost considerably less than graphical editors, probably due to the development costs of the different interfaces.

On the other hand, text editors do carry some negatives. Perhaps the biggest difficulty Web creation via text editor involves knowledge of good coding. Although a good knowledge of HTML can benefit you with nearly any program, if you do not have a strong understanding of HTML, you're likely to encounter considerable frustration using one of these programs. Additionally, some text editors are just that—text editors, and you may have to open your browser separately to review your Web site—a separate step that might produce, at worst, a slight inconvenience.

BBEdit

Macintosh-based Web creators who prefer a text-based design program frequently choose BBEdit, from Bare Bones Software, Inc. This high-performance HTML and text editor for the Mac includes a variety of special features, such as HTML shortcut tools, multiple undo, a built-in FTP function, and support for over 15 programming languages.

The main screen can host several open windows at a time, enabling users to review the document's contents and set display and file options.

While working in this display, users can take advantage of the special colored tags, to more easily determine the difference between different types of codes. Users can also review the extensive glossary of HTML tags, scripts, Web-safe colors, and special HTML tools. Another great enhancement window features a comprehensive search function that allows you to conduct multi-file or batch search operations.

Bare Bones Software, Inc.
www.barebones.com

This site features a variety of software produced by Bare Bones. Special features include product information, support, and updates for BBEdit. Additionally, through this site you can buy, demo, and/or register the program, or download BBEdit Lite, a toned-down, free version of BBEdit.

BBEdit Tips and Tricks
http://any.browser.org/bbedit/

Tips and tricks abound on this informative site, which highlights general tips such as creating your own extensions and using the FTP tool. More specific tips about BBEdit reside in the HTML and Perl tips section, or you can look into companion tools for the program.

HomeSite

For Windows users, HomeSite usually leads as the text editor (or self-proclaimed "WYSIWYN—What You See Is What You Need" editor) of choice. This sophisticated editor from Allaire Corporation provides users with a variety of helpful tools, including HTML shortcuts and templates, a great FTP function, HTML syntax checking, and Cascading Style Sheet (CSS) integration capabilities.

Starting with an interactive interface that lets you design your Web sites from scratch or a pre-designed template, try out the pre-generated shortcuts created by HomeSite, create your own shortcuts, or code your own HTML. HomeSite also uses colored tags to help you more easily identify different tags, and its easily configured menus, tabs, and assorted icons offer a high level of control over the content, usability, and interactivity of your site. After you've finished your site, take notice of HomeSite's extensive library of Web site checking tools, such as link verification, performance testing, spellcheck, and syntex and tag inspectors.

Allaire: HomeSite
www.allaire.com/Products/HomeSite
This marketing site provides users with a better understanding of HomeSite. Learn about HomeSite's new enhancements, development plans, and press coverage, or look into current information about obtaining the latest update. The site also overviews the product quite thoroughly.

Marjolein's Help for HomeSite Users
www.hshelp.com
This excellent resource quickly and easily helps you find the answers to burning questions about HomeSite. Learn about working with HTML, tips regarding the program's upgrades, relevant books, tools, and information about a variety of other functions of the program.

Notepad

Notepad, a popular text editor built into the Microsoft Windows operating system, offers a basic way to edit your HTML documents. Although not designed specifically for HTML coding, Notepad's uncluttered and intuitive user interface—just two scroll bars and a menu bar—is extremely easy to use and lets you quickly and effectively edit your HTML. Designed to produce a small application footprint, Notepad uses very little memory and disk space. Whereas some HTML editors will pad your HTML with additional and sometimes unnecessary code, Notepad keeps the code the way you write it. As long as you remember that Notepad is neither a robust word processing program nor an HTML editor, but a simple way to view and edit text, your success will be limited only by your HTML coding skills. Other than that, since Notepad comes built into Microsoft Windows, the price is right!.

Notepad
www.notepad.org

This site is a useful resource for product information on Notepad. Organized in a simple, straightforward manner, you'll find information here on key features as well as system requirements for loading it.

VI and Emacs

As diehard programmers who have been around since the onset of the World Wide Web and before, Unix fans turn to VI and Emacs. Both programs boast an avid following; users usually prefer one to the other and will heartily defend their choice. These "visual" editors offer users the opportunity to see a document as it is being edited—a fairly standard process these days, but one that helped drive initial interest to the program. Unlike some other Web editors, they also allow you to edit your Web pages directly on the Web server.

Essentially, the user opens either program and manually codes all of the HTML. The file is then saved as a Web page. VI and Emacs differ

> **Application Footprint**—*the amount of memory and disk space taken up by an application*

from the Mac and Windows programs listed previously because they are completely text-based, without any of the "pretty" graphical formatting presented by the other programs. Although they are extremely customizable and useful once you understand them, many beginning users may become frustrated by the technical brainset needed to work in the Unix environment. Before using either program, take a few minutes to learn some of the necessary commands; the programs will prove far more useful if you understand their basics.

The Unix Reference Desk
www.geek-girl.com/unix.html
Your complete resource for information about all things Unix-related, this site provides a lot of information about surviving in the Unix environment, including information on Emacs and VI. The site keeps true to the Unix look and feel, providing a simple, direct, technical series of explanations about issues related to the Unix operating system.

The VI Lovers Home Page
www.thomer.com/thomer/vi/vi.html
This site provides everything from versioning information to macros. Over 15 manuals and tutorials make this site invaluable to people struggling in VI, providing great information that the site owner cleverly ranks and describes. The site offers a variety of other helpful links, including FAQs.

The GNU Emacs page
www.gnu.org/software/emacs/
The GNU Emacs page features a variety of helpful information regarding the GNU version of the Emacs editor. Not only will the site direct you to resources from which you can download GNU Emacs, it also provides different types of help for using GNU Emacs, including manuals and mailing lists.

WYSIWYG, If You've Got
a Little Cash to Burn

Pronounced wizzy-wig, WYSIWYG stands for "What You See Is What You Get." Originally, the phrase referred to word processing and desktop publishing programs, but has since extended to include a variety of other programs, including those used for the development of Web sites. A WYSIWYG application lets you view, while you are working, how the document will appear in final output version. This differs, for example, from text editors that do not display different fonts and graphics on the display screen, though the formatting codes have been inserted into the file. Initially developed to eradicate the need to code in HTML, many WYSIWYG editors today seek to help Web site creators quickly and easily create Web sites, while still providing access to the HTML coding.

Just like creating a newsletter in WordPerfect or PageMaker, WYSIWYG editors let you drag, drop, and format graphics and other elements directly on the page, while the software codes the HTML in the background. Many WYSIWYG packages use toolbars and buttons similar to those used in popular word processing programs so that users can more easily identify the elements needed to create a high quality Web site. Many also offer templates and wizards to help novices (or those wishing for a little more creative assistance) with a quickly and easily designed Web site. Additionally, most incorporate a text editor to enable users to check the basic code, a definite hint that knowing HTML basics can give the Web designer a real advantage!

WYSIWYG opponents frequently voice their disdain for these programs, based on a variety of issues. One of the most difficult problems to work around involves the actual coding process. With straight text editors, users can be fairly sure their code will remain unchanged by the program; many WYSIWYG programs add extra coding or automatically modify even hand-coded text, causing frustration for HTML purists. Also, because of the multiple levels of programming necessary to create these software packages, the costs can easily range in the hundreds of dollars.

However, if you want to get your Web site up and running in a hurry, WYSIWYG programs provide a quick and easy option—and some can be acquired on the Web for free or a nominal fee. Especially for beginners, these programs feature an alternative to learning HTML and advanced programming. Aspiring programmers can also use them as a learning tool, by reviewing the coding pages during a page design process. Top WYSI-WYG tools include FrontPage, Dreamweaver, GoLive, and Netscape, although many additional tools exist on the Web.

FrontPage

FrontPage presents a high-powered tool for those of you who don't count yourselves among the ranks of the programmers or experienced designers. You don't need to know HTML, as FrontPage creates the HTML code for you. If you can use Microsoft Word, you can use FrontPage to develop a very respectable Web site. FrontPage essentially takes you step-by-step through the Web site development process, while allowing you to customize your site through a stash of templates, styles, and backgrounds. With more than 60 customizable pre-designed themes, it's fast and simple to produce a professional-looking Web site.

FrontPage does a lot more than just help you create your Web page. If you want to add a touch of Flash to your Web site, FrontPage 2000 Dynamic HTML (DHTML) enables you to add animation effects. FrontPage also includes several components that help you to manage a Web site (FrontPage Explorer), and even helps you turn a PC into a Web server (Personal Web Server).

When you first bring up FrontPage, you will see a blank page ready for you to start typing or adding images. Tabs allow you to change easily between HTML, WYSIWYG (the default view), and preview modes. The Views bar displays icons for performing various Web-related tasks, such as organizing your files and folders, and once you've created your Web site, you can analyze and review it with FrontPage's reports function. Easily add navigation buttons to your Web

Flash—*animation technology that runs via Web browser plug-ins*

pages through the Navigation view, and get a better grasp on your hyperlinks using the Hyperlink view. The Task view will help you to track and manage your Web tasks. If you're familiar with MS Word, you'll find FrontPage's editing toolbars to be completely intuitive, with the same formatting and editing icons that you're used to seeing.

In edit mode, you can add or modify text just like a word processing program, add images, create links, bookmarks, and hotspots. You can also insert Web components such as hover buttons, marquees, and hit counters. Jazz up your Web pages by selecting from a list of background styles and colors or applying a particular theme to your Web site. To really utilize the full functionality of FrontPage, such as forms handling and database support, make sure that your Web server supports FrontPage extensions.

Microsoft FrontPage
www.microsoft.com/frontpage

This is Microsoft's official site for FrontPage and contains specific and detailed information on the program's capabilities and features. Other highlights include details about using FrontPage, tips and tricks, other resources, and download and trial information.

FrontPage World
www.frontpageworld.com

A great resource for FrontPage, this site contains product information, as well as forums, training, downloads, reviews, and many other FrontPage-related links. You can join in the Tips and Tricks forum or suggest other great links for FrontPage to add to this site.

Dreamweaver

Macromedia's powerhouse design program currently leads the market for WYSIWYG products, and offers users the ability to work in a dynamic, visual layout. You can also opt to work with the integrated BBEdit or HomeSite programs in a text-editing environment. Or, if you prefer, you can view both at the same time, via a special split-screen function. The programmers at Macromedia have worked hard to produce a program that changes little of the original coding, and, for the most part, have succeeded via their "Roundtrip HTML," which preserves hand-typed or third-party tags.

The rich layout view enables users to create and edit tables quickly, or drag-and-drop images and layers. If you make a mistake, don't worry—the extensive History Panel helps users visualize workflow and quickly undo any procedure. Another super option is the "dream templates" feature, which allows content suppliers to directly edit only the content within a page, rather than risk compromising the site's design. Additional hot features include great integration with Fireworks and Flash, creating a dynamic editing tool.

If you're concerned about making mistakes, the JavaScript debugger and HTML cleanup functions will help guide you to a successful Web site. Don't miss the integrated information from the O'Reilly publications, which includes reference material on JavaScript, HTML, and CSS, as well as other helpful information. Also, since Dreamweaver supports W3C standards on HTML, CSS, and accessibility for people with disabilities, you can also check your page to ensure that your source code meets these standards.

Macromedia's Dreamweaver site

www.macromedia.com/software/dreamweaver

The complete resource for product information about its Dreamweaver program, this site discloses extensive, well-presented information. Beyond the new features, system requirements, and upgrade guide, check out the support and training areas. You'll also find links for online forums and Macromedia Edge, a customizable newsletter.

Dreamweaver Depot

www.andrewwooldridge.com/dreamweaver

Dreamweaver Depot provides numerous Dreamweaver extensions, from actions and commands to custom tags, themes, and inspectors. The site also reveals an interactive discussion area, news, and links to other great Dreamweaver resources.

GoLive

Adobe's answer to Dreamweaver, GoLive issues an excellent challenge to the industry leader through some unique and innovative tools. Its 360Code™ function claims to preserve all code formatting, along with simultaneous layout/source views, and customizable browser-set based syntax checking. GoLive's HTML source editor enhances this usability by coloring HTML tags and syntax, customizing text macros, and allowing complete control over the HTML composition and editing. Another nice tool, the Outline Editor, lets you see the source code in an outline form, rather than in straight HTML—this layered look can help you quickly troubleshoot your code.

Excellent integration with Adobe's Photoshop enables you to directly edit and optimize Web images without leaving the GoLive environment—this includes working with layers! You can also work with native Photoshop, Illustrator, and ImageReady files, making a quick change from GoLive without having to open the other files. If you're concerned about the load time of a page heavy with graphics, GoLive's load time checker will display estimated download times for a Web page.

Some of the best features of GoLive revolve around its site management and design functions. Live link-checking management helps you validate external links and JavaScript-embedded URLs inside of GoLive Actions and Rollovers. You can easily work through a network connection as you dynamically update files via FTP and keep a constant eye on the status of your network connection. GoLive also allows users to determine what files have been modified or created since the last upload to the Web server, so you can quickly and easily select which files to transfer. Additionally, through the integration of Microsoft's Web Distributed Authoring and Versioning (WebDAV), multiple users will no longer overwrite each other's changes.

Adobe's GoLive site

www.adobe.com/products/golive

Not just a product marketing site, Adobe's GoLive offers a plethora of useful information. Browse reviews in trade magazines and links to support and training assistance, or take a tour through a gallery that shows outside pages created in GoLive and those that highlight the program's special features. You can even browse through a list of related products, including a variety of publications.

GoLive Heaven

www.goliveheaven.com

As a great resource for GoLive addicts and newbies to the program, GoLive Heaven gives you answers you seek, from cascading style sheets to pop-up menus, via great demonstrations. This site also provides relevant articles for Web designers, files, and much, much more.

Build It With a Browser—Netscape Makes It Easy

You don't actually need any complex software at all to create a basic Web Site. There is a free, basic editor packaged with the full edition of Netscape—Netscape's Composer. This WYSIWYG (What You See Is What You Get) Web page editor truly displays your Web page as it will look to your Web site visitors. You will be able to create your Web site as easily as you would create a Word document. If you don't already have Netscape Communicator, you can download it for free from Netscape's Web site: http://home.netscape.com

- Just open a new, blank page in Composer and start to enter text and graphics in much the same way as you would in a word processing program like Microsoft's Word. Composer adds the required HTML tags behind the scenes. Toolbars and menu options let you format the text, insert images, and add background and text colors, etc. Whenever you're ready to see what your Web page looks like in the browser, use the "Preview" button on the toolbar. The user-friendly toolbar icons enable you to switch easily between preview and edit views, as well as print and publish your page.

♦ If you're using Netscape 4.7 or below, you can select from a comprehensive list of Netscape templates, download one, and edit it to create your own, customized Web page. In order to use this feature, you do need to be connected to the Internet, since you're selecting from the templates on Netscape's site. The collection of templates include a personal Web page, business formats, and resume pages.

♦ With Netscape 4.7 or below, you can use the Netscape Wizard feature. A Wizard is a program that takes you step-by-step through some otherwise complicated procedures. Even easier than using a template, the Netscape Wizard takes all the pain out of building a Web site You must be connected to the Internet and will be required to register for a home site with Netscape Web Sites. Once you've registered, you'll be allocated a URL for your Web site and will be led through the steps to customize it. Just follow the instructions—it's as simple as that!

Don't forget to click on the Help option at the top of the screen—it holds an index of information vital to successfully creating a Web page using the Netscape editor. Netscape 6 has a particularly excellent help section. Additional helpful resources include:

Netscape
www.netscape.com
Netscape's own Web site is essentially what you're used to seeing when you bring up the browser and has the usual available features such as downloads, instant messaging,

mail, and calendar. The site also provides technical support hotlines, resources, and a comprehensive list of Netscape services in a simple-to-use layout.

The Netscape Unofficial FAQ
www.ufaq.org/

Find the answers to almost all your questions about Netscape. This well-designed site includes information about upcoming versions of Netscape as well as the current. You can download files and release notes, and get information on the browser plug-ins.

Silly Dog Netscape Browser Archive
http://sillydog.webhanger.com/narchive/

A sort of "one-stop shopping" location for Netscape versions, this one-page resource is packed with information on just about every version of Netscape—364 are listed—with downloads galore. If you need help, there's a message forum and feedback form. Pictures of the various versions of Netscape are posted for the curious, as well as reviews.

Super Click and Easy

Maybe you don't want to think about creating your Web site? Fortunately, the Web can solve this dilemma for you. Depending on your needs, you can visit numerous resources for pre-designed templates—you only need to determine what text and images to add and apply them to your chosen template! These resources come in a variety of shapes and sizes—from sites that create a "take it and go" Web page for which you must find a host, to those that help you generate a site and host it for free, so long as you endure advertising linked to or placed on your site, to more professional tools that charge a fee for their template services and hosting.

ImageCafe

www.imagecafe.com

Check out the professionally designed templates for free—they cover many topics, from Finance & Money and Media & Communication to Religious & Organizations and Family, Recreation, Events, and more. If you get stuck, you can enter a live chat session, during certain hours, with a technical support representative. To keep your site, choose from a selection of reasonable payment options, some of which include hosting.

Perhaps you purchased this book to learn how to communicate with Web designers and programmers. If you don't plan to design your own business or personal Web site, this book will help you understand the different bits and pieces necessary to create a great Web site and enhance your appreciation of the people who do the work! If you're seeking a good designer or company to create your page, most likely you'll be overwhelmed by the variety of people willing to help you. Just treat the design of your Web site like any other work for hire: Look for an individual or company with a design portfolio you like and that won't break your budget—because once you get started planning your Web site, it's difficult to stop. Most importantly, if they can't discuss the concepts in this book with you, you probably should look for a designer with more experience.

Image Editors

Optimizing and perfecting images usually requires time, patience, creativity, and a little technical know-how. If you're lucky enough to have worked with a scanner, clipart, and an image editor prior to purchasing this book, congratulations. If not, you will find excellent information in Chapter 6, Your Best Image, about the ins and outs of image design and editing.

Don't lose hope! Today's software marketplace offers some excellent tools to help you create high quality images more quickly and easily than ever. Some of these resources will cost you both time to learn how to successfully work with the programs and money to purchase them. Others contain fewer bells and whistles and cost less; however, they suitably get the job done. Popular higher-end graphic editors include Paint Shop Pro, Photoshop, Image Ready, and Fireworks. Users frequently form attachments to their program of choice, but most of these programs meet the same purpose and beyond: the creation of quality, professional-level Web graphics.

Image editors are among the most frequently underused programs. Many people focus on changing the size of their graphics (both dimensions and file size) and forget about the other wonderful tasks these programs can perform. Not only can you create crisper, clearer images for the Web; you can also correct damaged photographs, brighten dark pictures, darken extremely light pictures, and turn a regular image into an artistic masterpiece unrecognizable as the original image. Get lost in image creation and manipulation through the following exciting image editors:

Paint Shop Pro

Paint Shop Pro, an image-editing tool with a set of robust printing and drawing capabilities, helps you create new images for your Web document, or convert and edit images that you've acquired from elsewhere. Although not entirely intuitive for beginners to use, Paint Shop Pro boasts an impressive array of features and supports more than 40 image formats. Don't miss its extensive special effects (over 75!), which include dockable toolbars, built-in special effects filters, RGBB color separation, masking options, multilevel undo functionality, complete layer support, "picture tube" brushes, CMYK separation, pressure-sensitive tablet support, flexible printing and retouching brushes, and adjustable cropping and selection tools.

Advanced graphic functions in Paint Shop Pro enable you to create high quality, animated GIFs, great visual mouse-overs, and image maps to add flare to your site. Speed up your download time through enhanced optimization and image-slicing options, and preview the results of your efforts quickly and easily through the top browsers. For an inexpensive graphics program, Paint Shop Pro helps you decrease your confusion and increase your productivity through tools like grids, guides, alignment, and grouping.

Jasc software
www.jasc.com
This attractive and well-organized site gives you a wealth of information on Paint Shop Pro, with downloads, reviews, and links to other resources. A particularly useful feature is the Jasc Software Learning Center, which has numerous tutorials and an online chat forum.

Lori's Web Graphics
http://loriweb.pair.com
First, stop and read up on some of the basics of Paint Shop Pro, then move on to the well-structured and easy-to-follow tutorials found on this site. An excellent links section gives you many great sources of graphics info, and there's even a "goodies" section, with a collection of original graphics for download.

Photoshop/ImageReady

As one of the leaders in the image manipulation program marketplace, Photoshop provides more than enough attributes for any aspiring Web designer—some people never get beyond its cropping, resizing, or optimizing capabilities! Adobe's premier image tool creates professional-level images, with a little bit of work on your part. Once you increase your comfort level with the program, usually by "playing" with different graphic effects, you'll be surprised at how quickly you can move to advanced functions; but don't expect to get to all of the functions—Photoshop includes over 100 types of filters!

With special vector editing and layering tools, you can quickly and easily create dynamite buttons and images, and three-dimensionalize those images with a large assortment of bevel and drop shadow effects. If you want to resize or distort an image to meet a specific purpose, Photoshop's liquefy command helps you interactively push and pull pixels in your images. For Web images, you also can ensure that you work with Web-safe colors, or if you forget to turn that feature on, you can "snap" existing colors to Web-safe colors.

Nearly anything can be accomplished with Photoshop and its bundled version of ImageReady, a tool well-suited for advanced Web image tasks. Slice large images into smaller pieces to ensure quicker load times. Uncomplicate the creation of image maps, rollovers, and animation to produce a higher-tech set of images for your site. Most importantly, take all of the steps necessary to output a high quality image with a very small file size.

Adobe's Photoshop site
www.adobe.com/products/photoshop

First, take a moment to peruse the numerous functions of Photoshop, then move into the impressive gallery of graphics created using Photoshop. But the bonuses don't stop there—this excellent site also provides step-by-step tutorials; quick tips; downloadable goodies, updates, patches, and plug-ins; an extensive, searchable support database; and more.

Photoshop Paradise
www.desktoppublishing.com/photoshop.html

Wow! Photoshop Paradise gives you a large assortment of links to useful tools, including actions and filters to download. You can also access a great assortment of tips and techniques, as well as links for a variety of other online resources related to Photoshop.

Fireworks

Macromedia's Fireworks, Adobe Photoshop's strongest competitor, provides a sophisticated Web graphics program that enables you to create, edit, and animate Web graphics that can then be exported to other products, such as Macromedia's Dreamweaver and other HTML editors. Aimed at Web developers, its user interface allows you to work in Flash, Fireworks, or Dreamweaver. Features include drag-and-drop rollovers, live animation, pop-up menu creator, refined Photoshop import-export, masking and layers panel enhancements, roundtrip table editing with Dreamweaver, and selective JPG compression.

You can easily manipulate the painting tools, which have drag control handles and points; they also contain a multitude of styles, including oil brush, charcoal, and watercolor. Fireworks lets you quickly and easily modify effects at any stage and update them on the screen. Its extensive palette features offer paint-style brush strokes, texture fills, and plenty of other special effects. For higher-end Web sites, you can implement the interactive image map tools, which help you create server-side image maps; you can also assign URLs to your drawings with the URL toolbar.

Macromedia
www.macromedia.com/software/fireworks
This is the main product site for Macromedia's Fireworks. Its link menu on the side of the page includes a wealth of informative resources, including product info, reviews and awards, support and training, forums and newsletters.

eHandsOn
www.ehandson.com
This well-designed site contains training on a selection of products that includes Fireworks. See Lisa Lopuck's Fireworks Fundamentals course; although it's not free, you can enroll at a reasonable price. To try Fireworks out for free,

request the sampler package of three—Fireworks 4, Flash 5, and Dreamweaver 3. This site is constantly updated so you'll always get up-to-the-minute information.

Illustrator

Adobe's Illustrator, the true designer's tool, produces high quality graphics that translate well for Web site use. Unlike the other graphics software mentioned in this chapter, Illustrator is intended for the creation of graphics, as opposed to the manipulation of graphics. It enables you to control every aspect of the graphic—to "draw" on screen—rather than limiting you to a smaller range of options. However, because of Illustrator's capabilities, it may require a bit more learning time, and the program does run in a higher cost range. If you can overlook those aspects, you won't be disappointed with the results, which you can quickly and easily import into other Adobe programs helpful to your Web design efforts.

Specific advantages of the Illustrator program include the ability to edit drop shadows, create a convincing glow around graphics, and more. The type controls let you manipulate text beyond your expectations—creating fantastic graphics you can use for your Web site. Illustrator enables you to overcome troubles with transparency, creating clean graphics with smooth edges. You can also use Illustrator to lay out a full Web page, produce graphics for animations, and optimize your graphics.

Adobe's Illustrator site
www.adobe.com/illustrator

Adobe's robust Illustrator site provides an excellent starting point through which you can learn more about the product and gain some special tips and tricks for its use. Check out this site for a gallery that highlights the full capabilities and special features of Illustrator, as well as an assortment of tips, user forums, and easily accessed updates.

Illustrator Resources
www.illustrator-resources.com

This easily navigated site offers many fantastic resources for your Illustrator needs. Great resources include downloadable brushes and plug-ins, as well as a section called "inspiration," which provides expert creations for your viewing pleasure. Check out the great tutorials, where you can learn about everything from brushes and filters, to masking and paths, and more!

3

Good Design Basics

This book seeks to point out the numerous achievements in the various fields associated with the creation of Web sites. Beyond the examples within the individual chapters—from the tools used to create the sites, content design, graphic design, coding, and more—you need to be aware of some of the essentials of good design basics. Expert opinions vary on the difference between good and bad design effects, but one statement remains true: If you build an easy-to-use Web site that provides dynamic, quality content, you will encourage users to return. That's the secret behind the success or failure of Web sites.

The Web itself provides one of the best teachers for good and bad examples of look, feel, and mobility, and its community of novice to expert users will quickly advise you of the quality of your site. Therefore, it is important that you spend time looking through different Web sites to determine what appeals to you, keeping in mind that, although "beauty is in the eye of the beholder," a Web site only achieves success through a successful combination of look, feel, and navigation. You can also utilize the expertise of the networking resources listed in Chapter 1, Getting Started. You can submit concepts, mock-ups, and questions to these groups, and, due to the nature of the Web, rest fairly well assured that those users will let you know exactly what they think.

Ironing Out Web Usability Kinks

The penultimate step to good usability requires you to define your goals and expectations. Will your site serve to sell your products to middle-aged, single businesswomen whose native tongue is not English, or will it provide a useful research tool for engineering students who want to learn about mechanical theory? Before proceeding with a site, involve all of the decision makers and gain consensus regarding the goals—otherwise, your site will be off to a rough start. The following books provide great resources that will assist your search for Web perfection:

- *Designing Web Usability,* by Jakob Nielsen

- *Designing Highly Usable Web Sites,* by Tom Brinck, Darren Gergle, and Scott Wood

- *Don't Make Me Think,* by Steve Krug

Nearly everyone on the Web has a different idea about usability, graphic user interface (GUI), information architecture, etc. Take a moment to look through these great collections of information that get right to the point and show you a variety of fresh and proven ideas related to good design basics. They also provide excellent examples and links to other great resources you won't want to miss.

Jakob Nielsen's site
(Usable Information Technology)
www.useit.com

Nielsen, known as one of the best sources for Web usability issues, guides designers around the perils and pitfalls encountered when creating great sites. Alertbox, the bi-weekly column on Web usability, proffers some true gems in the topic area, including analyses of sites that fail usability standards, reviews of why new technologies work—or don't, and many other gems.

Usability Professionals Association
www.upassoc.org

The official Web site of the UPA, this excellent resource will keep you up to date with all the current industry conferences, outreach projects and activities, tools, articles, and progress of usability issues. There's an extensive resource list of organizations associated with ergonomics and human-computer interaction, as well as a good selection of usability style guides.

Additionally, know your audience. When determining demographics to describe the people who will visit your site, don't choose the easy way out and simply list "college students" as your audience. Instead, research your brick-and-mortar customers to provide specific demographics, such as:

- Age—can help you determine engagement level

- Technical experience—can affect graphic and jargon choices

- Education level—may define reading level of your content

- Industry—helps identify interesting topics

- Location—perhaps some geographic data will add more value

- Language skills—affects the creation of your content

- Income—bundled packages or special deals may be more important

The following tips provide information that you can find in other chapters within this book. They highlight the important factors to consider when creating a site:

Top Ten Essential Web Design Elements

1) Content includes interesting, useful information: No one likes a boring site. If the content within a site contains too much technical jargon, long-winded, overblown language, or leads readers around in circles, chances are greater that users will not return. Chapter 4, Content to Keep 'Em Coming Back, contains more information to help you ensure that your content continues to drive users back to your site, rather than driving them away.

2) Effective navigation enhances visitors' use of the site, rather than detracting from it: The best content and graphics won't help if site visitors cannot figure out how to find the resources. When you successfully use navigation tools, you enhance the user's experience, which correlates to your site's effect. This chapter will investigate how to work with the devices that make site navigation work for you and your users.

3) Printing your site provides pleasure, not pain: Some Web designers get so caught up in the effects and beauty of their site online, they forget about printing issues. Don't fall into this trap. People still enjoy printing out information and taking it with them to read later—on the train, plane, or when their eyes need a break from the computer. This chapter will offer some insights on creating successful printouts.

4) Pages on the site load quickly: Despite the fact that computers, Web browsers, and Internet connection options continue to increase, some people still suffer connection speeds of 56k or less. Don't take the chance that elements of your site will eat up precious bandwidth. Check out Chapter 8, Special Effects, to determine whether your site meets optimization standards.

5) Technology is leading edge, not bleeding edge: Don't fall so in love with the latest technology, whether cascading style sheets, new programming languages, or new applications, without first ensuring they work across all platforms. After all, if only 5% of the Internet population can see certain "bells and whistles," why not wait a little longer before implementing them?

6) Use of graphics results in quality, not quantity, and does not overpower site visitors: Sites that contain 15 animated gifs and 30 other graphics don't usually impress your visitors; they often scare them away! Learn how to successfully create quality, low-bandwidth graphics and work with animated images (Chapter 6, Your Best Image) and with colors and alignment to create a graphically appealing site.

7) Site updates frequently, so users can trust the information: Have you visited sites that reference information ten years old or show their last update as three years ago? How well did you trust the information on that site? Encourage your users that you are providing accurate, up-to-date information by updating your site frequently. More information on how to do this can be found in Chapter 4, Content to Keep 'Em Coming Back.

8) Site has clear objective, both for audience and in presentation: When you determine the initial reason for creating a Web site, make sure you work your entire site around that reason, then step back, and determine whether your audience will seek your site for that reason. If not, you'll encounter some confused users. By creating content that mirrors your site mission, you can avoid this problem.

9) Alternative methods of navigation include search engine and site map: No matter how perfect your navigation structure, you'll always encounter users who just can't figure out how to move around within your site. Make their lives easier and support emails fewer and further between by providing alternative means of navigation. This chapter examines different methods for achieving that goal.

10) Keep informed about Web standards: Read Web design magazines, subscribe to listservs, and network to find out the latest and greatest standards for Web design. Then, research that information through other sources to ensure you know the true standards, rather than the fads that do not stretch across all platforms. Consider information about monitor size, number of colors, bandwidth, and anything else you can find.

All About Layout—Placement is Everything

By now, you probably cannot wait to launch your site—after all, you know about and may have already acquired the design tools you need, you know why you want to build a site, and you have identified the type of site you plan to build. So what can possibly go wrong at this stage? Plenty. The layout of your Web site and the placement of objects on your pages can make or break your Web site. Just as a successful interior designer will arrange a roomful of furniture and soft furnishings so effectively that everyone drools, careful attention to detail in the strategic arrangement of your text and graphics can make an outstanding, and lasting, impression on visitors to your Web site.

The Fight for Front Page Real Estate

Three basic issues provide challenges to Web "real estate" battles. First, if you work with clients, they'll want everything on the front page and it will look hideous. When working in a corporation, everyone wants a piece of the pie, and you must diplomatically balance their needs and your design knowledge. Finally, when working by yourself on a private site, it's hard to figure out what sections should get the highest notice.

Place logos and other frequently used graphics, such as navigation bars, in the same place on each page so that your visitors feel comfortable with the navigating your site. Your site logo should serve as the beacon that consistently shows your identity and provides users with a way to navigate to your home page. More information on this will follow in the navigation section of this chapter.

Aligning for Success!

The overall look and feel of your Web site will depend on many things, including the alignment of text and graphics—how your Web page elements line up relative to each other. Poor alignment is a common problem seen on many Web pages, and when you don't get it right, your Web page will look messy and unprofessional. In particular, if your pages contain a lot of information, you must pay special attention to the layout. This not only will produce a cleaner-looking site, but also will ensure better communication with your visitors. It's tiring for the eyes to have to wander aimlessly about the page without any anchors that help the information make sense; as a result, you will lose many visitors. Sites like the U.S. Fish and Wildlife Service (www.fws.gov) and National Wildlife Federation (www.nwf.org) effectively portray good examples of successful alignment of text and graphics. Take a look at them and browse through other pages on the Web to determine which alignment styles please your eyes the most.

Generally speaking, we're used to seeing text left-aligned, which means that the text is placed flush against the left margin or against an invisible vertical line. Keep this in mind when you decide how to align your text; right-aligned text tends to look odd, and centered text should be used sparingly or it will become tiring to look at. Main headings and titles look good centered, but avoid centering large blocks of regular text, lists, and some items within tables, as this can make your page look disorganized and will weaken its impact. If visitors to your Web site have to work too hard to see everything, they will not stay. You want immediate impact and for the eye to find the layout pleasing and uncluttered. On the other hand, make sure you don't have your text crowding against the left edge—and as another tip for improving the overall look of the

**My Visit to Stirling Castle
in Scotland, 2001**

Example of poor alignment: neither the image labels nor images line up.

page, give your text a little breathing room through the effective use of white space.

Horizontal alignment is also extremely important. If, for example, you are creating a photo gallery, you should make sure that the bottoms of the pictures on a particular row line up along an invisible horizontal line known as the "baseline." This will help you to avoid having, for example, three similar objects such as labels, where one object has the text closer to the top of the page, another has the text toward the bottom, and another has the text that appears between the other two labels. Remember, as with everything, be consistent in your alignment so that your Web site has a polished look. A well-organized site is a well-aligned site!

Left/Right/Center

If you're using a software package to build your Web site, you'll find that moving your text and graphics around is a simple process of clicking on an icon or menu item, or even an "alignment" button. If you're writing your own HTML code, a few additional tags or attributes will give you the power to take control of arranging your Web objects as you please. Take a look at Chapter 5, Code Challenges, for more information on tags.

You can use either the <CENTER> tag or the ALIGN attribute with other tags such as <P>, <IMG>, <H1>, <HR>, and so on. For example: <HR ALIGN=LEFT> aligns a horizontal line with the left margin. A word of warning: Some browsers don't support all HTML tags and sometimes they will ignore all centered and right-aligned text and graphics and will interpret them as left-aligned only. Don't go crazy switching between text alignment styles—find a style and stick with it!

white space (negative space): *the open space between design elements (graphics, text, etc.) that increases readability*

HTML Tag or Attribute	Description and Example
<CENTER>headings, text, and graphics that you want to be centered go here</CENTER>	Tag that surrounds text and graphics to be centered.
ALIGN="RIGHT"	Attribute that aligns text or graphic right
ALIGN="LEFT"	Attribute that aligns text or graphic left
ALIGN="CENTER"	Attribute that centers text or graphic

Text Layout Tips

Restrict your lines to contain about 40-60 characters and your paragraphs to 4-8 lines to ensure effective readability. These tasteful tidbits produce more eye-appealing results for your visitors. You can use tables and style sheets to more creatively arrange your text so that it breaks up lengthy blocks of text. The average visitor does not read online in the way he or she would read a book, but scans the page on his or her monitor to see if there's something of interest. Therefore, keep your text blocks relatively short so that you can display your information in "chunks."

Vary your text styles to produce a more inviting landscape to site visitors. By creatively integrating bullets, sidebar, and headline elements, you can attract and hold the attention of the Web surfer.

Image Placement

When preparing to add images to your Web page, take some time to think about where you would like to place them with respect to the text on the page. In general, the text is the crux of your content, and usually the most important element of your Web site. The images can really add value to the text by complementing it, drawing attention to it, and essentially making its overall presentation appear pleasing to the eye. They should never overwhelm the text (or the user) by being the most important element on the page, unless they clearly support the main focus of the page.

When working with Web pages, it helps to have an artistic flair for page layout, but by following one or two general guidelines, most people can put together an attractive combination of text and images. Where you place the image relative to the text is extremely important and can make or break the appearance of the Web page. Wrapping the text around the image can be extremely effective and make the image appear more integrated with the rest of the content. Space considerations may require you to place text immediately above, below, or next to the image—there are several options and the final decision will be yours. You can choose to position an image, relative to text, in one of five ways:

Top   The text is aligned with the top of the image.

Middle   The text is aligned with the middle of the image.

Bottom   The text is aligned with the bottom of the image.

Left   The image will appear on the left of the browser window and the text wraps around the right of the image.

Right   The image will appear on the right of the browser window and the text wraps around the left of the image.

Castle Urquhart
Text ALIGN="Top"

Castle Urquhart
Text ALIGN="Bottom"

Castle Urquhart
Text ALIGN="Middle"

This is what happens
when the text
is left-aligned:
<IMG SRC="urq1.jpg" ALIGN="LEFT">

This is what happens
when the text
is set for right
<IMG SRC="urq1.jpg" ALIGN="RIGHT">

Examples of text-image alignments

A Little Color Goes a Long Way

If Web sites were restricted to black and white display colors, the resulting World Wide Web would be very dull! Fortunately for us, the Internet practically glows in glorious color, and it only makes sense for you to take advantage of this when creating your Web site. Intelligent use of color can make the difference between a successful, energized, inviting site and a dull one that scares away visitors.

While complete abstinence from colors won't produce the successful Web site you desire, you'll generally find that the more professional and sophisticated Web sites opt for quality, not quantity, in color selection. If you're creating a corporate or product site, you may be restricted to using corporate or logo colors. If, however, you can exhibit a bit more color freedom, try to use colors to indicate the "personality" of your Web site. For example, colors like reds, yellows, and oranges generate a dynamic and exciting feel (www.matchbox.com), while blues and greens result in a more restful response (www.relax.com). Colors can also show structure— if you opt to make all of your first-level headers bright purple, and your body text remains black, the result will cause your headlines to gain even more prominent notice than their larger size and boldness.

Your text should contrast easily with the background color, especially since your viewers may modify their monitor hues, contrasts, and brightness levels, resulting in less contrast than on your computer. For example, avoid placing dark blue text on a purple background—the result could make your text virtually unreadable as the colors compete for notice! On the other hand, although a bright, pretty blue on a sunshine yellow background might jump off the screen at people, their eyes will soon become tired from the glare. You should also be careful when using light text on dark backgrounds, as this produces an effect difficult to read. Stick to neutral, dark text on a non-glaring, light background for the best effect. Poor color usage on Web sites is a common complaint among people who are color-blind or who have poor vision. For more information on addressing issues for people with visual disabilities, refer to Chapter 8, Troubleshooting Your Efforts.

Firelily Designs
www.firelily.com/opinions/color.html

For some detailed and technical information on color in design, you should take a look at this Web site which gives some really great insight into why colors work the way they do and how various components of colors combine to give you the actual color you see.

Pantone
www.pantone.com

Learn from the people who practically invented the color wheel and have been providing advice on color combinations since the early sixties. Not only can you find some neat products to help you with color matching, you should also check out their "All About Color" section, which provides invaluable insights into using colors.

Despite all of your valiant attempts to create a pièce de résistance, unless you are working with style sheets (see Chapter 5, Code Challenges, for more information), your color effects will revert to the default Web browser settings or modified browser settings. However, you can safely assume that people have not modified their browser defaults to that end. If you are not satisfied with the default color scheme assigned by the browser or your Web creation program, you can easily customize the color scheme for your Web site. The three standard page elements for which you can select colors include:

- Background—Choose a neutral background, especially if your text colors change. Be careful of using a textured or image background, as it may make your text unreadable.

> **Hexadecimal**—*a numbering system which, instead of using 0-9 as in the decimal (base 10) system, uses 0-9 and then A-F to represent 11-15 (base 16 system)*

- Text—For the best results, stick to black or other dark, Web-safe colors

- Links—Let your link colors change when visited, so your users can more easily determine the pages they have visited in your site.

Playing around with color schemes is not difficult, and if you're using HTML code, just a few basic commands can set you up with a great color scheme. Special color codes consisting of six-digit strings of letters and numbers (a hexadecimal number) tell the Web browser which color to use when displaying the Web site. The code is formatted as **#rrggbb**, where each part of the code (e.g., rr, gg, bb) represents a variation of one of the primary colors (red, green, blue) and the entire code represents a combination of the three primary colors. Simply insert the color codes into the appropriate HTML tag and attribute. For example, to change the background color, use the **BODY** tag with the **BGCOLOR** attribute: **<BODY BGCOLOR="#FFFF00">**.

If you prefer not to work with the rrggbb color codes, you may opt to use standard HTML color names, both of which are supported by both Internet Explorer (IE) and Netscape 3.0 and up. For example, instead of using the code **#FF69B4** you could use HotPink. However, this will certainly cause problems with visitors who are not using IE or Netscape, or who are using pre-3.0 versions of IE and Netscape. Note: If you decide to change the colors of text and links, be aware that you should assign different colors to Normal Text, Link Text, Active Link Text, Followed/Visited Link Text, and Background. For example, **<VLINK="#rrggbb">**. The default colors (and therefore, what most people are used to) for standard elements include:

Element	Default Color	Definition
Normal Text	Black	All text that is not a link
Link Text **<LINK>**	Blue	All links that are either not currently active, or followed/visited.
Active Link Text **<ALINK>**	Red	Just after a link has been clicked, it will change color to show that it has been "used."
Followed/Visited Text	Purple	Links that you have previously Link clicked on and activated.
Background	White	The entire space of the page that sits behind the text.

Browser-Safe Colors

You may have heard the term browser-safe or Web-safe colors. Most computer monitors can easily display at least 256 colors (8-bit color), but only 216 colors are really common to most browsers and operating systems. Normally, the browser used by the Web surfer will interpret the background and text colors displayed on the computer screen. For example, in Netscape Navigator, take a look at the Preferences dialog box and select Display Color—the settings are in the Color subcategory of Appearances. If a browser is not able to support a color, it deals with this in one of two ways. It will either change the requested color into a color that it is able to support and display (the closest possible color to the one requested), or it will "dither" the color, which means that it will combine two supported colors to approximate the color requested. Therefore, if you want to truly represent your color choice, choose Web-safe colors.

Color Charts

There are literally hundreds of color codes and it's very unlikely that you will remember, or even use, most of them. Color charts are available for you to select from, and there are many Web sites available that post color charts for you to use.

HEX Color Chart
www.interlog.com
This site offers an excellent color chart. You can easily see the codes next to the colors to ease and speed your color choice.

Super Color Chart
www.zspc.com/color
Check out this interactive color chart. To determine the code to use for colors you like, just click on the relevant color and see the code displayed.

Brobst Systems

www.brobstsystems.com/colors1.htm

This site has a nice color chart that shows not only the #rrggbb codes, but also the actual color names—just in case you prefer to use the names.

Color Disasters

If you choose the wrong combination of colors, your visitors will not be able to see your text, so be very careful with your color selections, or you will join the ranks of the numerous Web site color disasters. Color can work for or against the content of your site, so take it seriously. If, for example, you want your Web site to portray serenity and harmony, you should normally avoid garish reds and oranges since these are traditionally the colors of energy and dynamism. We don't want to really list any specific sites as examples of color disasters, but we're sure you'll find plenty by yourself!

A word of warning: Since the Internet is truly an international medium, if you plan to reach global audiences, you should investigate any color symbolism that various countries may have. For example, the color purple is often associated with regality (think of ancient Rome), spiritualism, mystery, or even death, depending on which part of the world you come from. Interesting sites to check out for color symbolism include:

Color Symbolism Chart

http://webdesign.about.com/compute/webdesign/library/
weekly/aa070400b.htm

This really great quick reference chart shows not only the general symbolism associated with colors but also shows color symbolism by different cultures.

Color Matters

www.colormatters.com

For more information on the importance on the right color scheme on the Web, this site is extremely informative on just about anything to do with color, including color symbolism. It highlights color as it relates to the eye, body, science, computers, and more, while also offering some dynamic communications tools through which users can further discuss color issues.

Common Causes of Color Chaos

- Black backgrounds
- Fluorescent text
- More than three main colors
- Light text on dark backgrounds
- Non-contrasting backgrounds and text

Creating a Clean, Well-Lit Site: Navigation

In an ideal Web creation environment, content managers work closely with graphic designers and Web programmers to create an interesting site that successfully incorporates graphic and text elements. Navigation serves three essential purposes: directing people to their destination, identifying the present location, and providing a return to the original location. All three intents can be accomplished through a variety of information architecture efforts, including successful site organization, a site map, and other navigational aids, such as crumbs.

Successful navigation begins with your front page, which should successfully introduce your users to the navigational scheme of your site, as well as provide a quick link to a site map as an alternative navigation method. If you decide to initiate navigation through the use of image maps on your front page, keep in mind that users who have switched off graphics on their browsers will be unable to see the image map; you should provide alternative methods for accessing the information on your site.

Site Outlines and Directory Structures

As you create your initial site outline, first identify the most important elements of your site—do you want visitors drawn first to your products and services, to investors' information, or both? How will your users seek information on your site? Identify the key categories, then the subcategories within those sections, and keep going until you reach the "bottom" pages. By "building out" the areas within your outline as you acquire and/or determine the need for additional content, you gain a better understanding of your site's organization, as well as missing components.

As you build out your site, keep in mind the directory structure for your site. The worst way to populate your site, and a rule frequently broken by new Web designers, occurs when you place all of your site files in the same directory. This results in confusion whenever you and other programmers add new files to your directories, and will cause frustration when you search through hundreds of files for the file you need to edit. Instead, try saving the top directory for the index page of your site. Then, you can separate your graphics, html, or advanced files several ways. Also, make sure that each directory includes as its main page a file named index.html. Otherwise, you could open your site up to security issues— not following that practice lets visitors pull up a list of all of the files in that directory, and could allow them to open files you do not want them to see. Some sample directory structures:

```
html
    pressroom
        media_releases
        executive_bios
    investors
        stock_ticker
        annual_reports
    employment
        job_listings
        benefits
graphics
    pressroom
    investors
    employment
cgi
    pressroom
    investors
    employment
```

```
pressroom
    graphics
    html
        media_releases
        executive_bios
    investors
        graphics
        html
            stock_ticker
            annual_reports
    employment
        graphics
        html
            job_listings
            benefits
cgi
    pressroom
    investors
    employment
```

You must also determine the naming convention you will use for your files, a process that frequently goes hand in hand with the directory structure process. To ensure that files may be more easily edited and found (and moved around, if misplaced in the wrong directory), consider starting all press releases with the initials pr. Then, you could follow with the date of each release (year-month-day), creating less confusion when new releases are added to your site, and resulting in press releases filenames that look like: pr010312 (press release dated 2001-March-12). Starting with the year, followed by the month and date, will create a clear advantage when you attempt to list the files by the press release date, since systems organize by numeric order. Since some systems are case sensitive (a file name that includes capital letters will be interpreted differently than one in all lowercase), play it safe by naming all of your files in lowercase characters. Also, avoid using misspelled, irreverent, or misleading titles; don't embarrass yourself by titling a bio "big_jerk.html" and thinking that no one will notice—because they will!

Finally, ensure that you use good title-naming protocols, since that can translate into the way search engines and dynamically generated site maps display results. Titles should consist of clearly understood words that concisely describe the page they represent. Starting all of your files with your company name, for example, will result in search engines internal and external to your site returning results of "Gryphon Communications." Although search engines will list the first few lines of text from your page as a description, the same repeated title will not enable users to easily navigate to the page they seek.

Site Maps

When Web site creators refer to site maps, they usually intend to describe a tool that quickly and clearly lays out the "organization" of your Web site. When prepared correctly, site maps provide a fast navigation tool for your site visitors—a clean, well-arranged Web page that breaks down the different sections and subsections of your site—similar to a table of contents. No matter how well you design the navigation for your site, your site map will always provide a fail-safe for your visitors. It should always be accessible from your home page, to enable users to find it quickly. Keep in mind that if you commit to offering your visitors this valuable tool, you must also ensure that the information will be consistently updated as your site changes, otherwise you might as well avoid including a site map!

After you apply your site outline and complete your Web site, you need to determine what type of site map to present to your audience. These important elements to your Web site serve a crucial purpose: If your users experience any difficulties finding material on your site, your site map quickly and easily outlines the structure of your site and points them on their way. You can use many Web design programs to automatically generate your site map, you can invest in a program specifically designed for this purpose (like Site Mapper, Xtreeme SiteXpert, or SurfMap, all available on www.tucows.com), or you can manually create your site map. Hand creation enhances your control over the site map, enabling you to determine the different categories you want the users to see, and the extent to which the subcategories break down. However, if you start by following a good structure, you can easily utilize dynamic processes to generate your site map.

You can limit the subdirectories that you display within either type of site map; however, including all directories up to the point of those directly above your files will provide your visitors with a richer, more complete view of your site. Additionally, although you can produce your site map as a graphic or series of graphics, you can save your users' bandwidth and your time by sticking with text-only site maps. Whether you hand-code or dynamically generate your site map, you'll need to determine the way in which you present the final information, either in an indexed or categorical format.

An indexed site map organizes your site's information in an alphabetized index list. Although this type of site map may seem inherently easy to the end user, it forces them to guess how you may have labeled your pages. Granted, if you title your pages well, this site map could enhance usability, but it's more difficult for end users to quickly and easily reach the pages they need. Look through the following examples for a basic understanding of indexed site maps: http://gbgm-umc.org, www.fifty-plus.net, and www.dnmyers.edu.

The other option, a categorical site map, results in organizing your site map in the same way you have organized your directory structure. This type of site map presents several clear advantages, including "teaching" your site visitors familiarity with your site's organization. When they select a link from within your site map, they see the structure of the directory(ies) it encompasses, which will aid in their navigation of that directory. Some demonstrations of different, successful categorical site maps include www.ebay.com, www.apple.com, and the extremely innovative site map at www.vw.com—note also that you can easily find the site map on the front page of both sites (with the exception of www.vw.com), one link down through the site info hyperlink. Here are two examples of different site maps:

Graphical Site Map

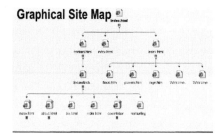

Categorical Site Map
Colors
  red
  green
Shapes
  circle
  square
Sizes
  big
  little

Linking: The Essence of Web Navigation

Often, the task of providing a successful link hierarchy falls on the shoulders of content managers—after all, they should hold the strongest familiarity with the text—however, programmers and designers can add value to this process, too. Linking can provide a great way for users to navigate within your site. An index at the top of a long page can connect different blocks of text within the same document, easing navigation for the reader (make sure you include a link to return to the top of the page, at the end of every section). Or, you can use hyperlinks to take your audience to a glossary that provides them with more information about complex terms. Additionally, successful link usage can serve to guide the readers through different sections of the site you wish them to follow, much like a flow chart.

Links must be understandable and diverse. Although "click here" frequently tells people exactly what to select if they wish to travel to another section, too many "click here" references result in boring content for your reader as well as accessibility issues (see Chapter 8, Troubleshooting Your Efforts). Instead of saying "To read the article about cats and dogs in the January 1 issue of *The Post*, click here," try hyperlinking the information to which you want to send users, through an example like this: "The January 1 *Post* article, "Cats and Dogs Stop Fighting."" In this case, the programmer would link the title of the article to the article, creating a more seamless navigation experience for the site visitor.

Keep in mind that hyperlinks emphasize the text they represent—they provide a good method of highlighting, but should be used sparingly. If you hyperlink every third word in your text, for example, the result would needlessly confuse readers. First, they would not know which information should be viewed as the most important. Second, they would spend a considerable amount of time clicking on all of the different links; rather than paying attention to the real subject matter, their attention would wander to all of the different links. You might cause them to lose interest, so pay attention to your links and make sure they are as carefully planned out as the rest of your site.

Crumbs: Leaving a Trail to Follow

No matter how carefully you plan out your site navigation and site map, you invariably will end up with a few lost and confused users. Fortunately, a special tool known as "crumbs" provides a trail that site visitors can easily follow—not just back to home, but also to higher level categories related to their current page. Crumbs provide information about the location of the user—offering a map that clearly states "you are here" and shows how that page relates to the rest of the site.

For example, a user reading a description of an item called "Skyscape" may encounter some frustration as he/she tries to determine the nature of the item. If the description is not complete and your site sells a variety of products, how will the user determine whether this is a record, painting, or book? Crumbs could produce the following result at the top of the page:

Home> books > religion

The user instantly knows that the item belongs not only in the books category, but is a book about religion. If the book does not meet your visitor's needs, perhaps another book in the "religion" subcategory will. The result improves the clarity of the different sections.

Crumbs can also take users to the next higher level in the site's hierarchy, letting them browse through similar items at that level. If a user has accessed a page via a search engine, for example, crumbs will show which category they're in, and possibly lead them to other resources in that category. Otherwise, unless the user clicks through the main organization of the site, they could miss out on some great material. Yahoo.com and About.com display an excellent method of using crumbs to improve usability, especially with the variety of information contained on their sites.

Navigation Bars

Navigation bars help your users navigate your site by providing a non-moving resource upon which users can rely. They should highlight the most important information on your site—the top categories to which users will try to navigate most frequently. Navigation bars appear in all shapes and sizes—from the tabular look of www.cancer.org and www.egghead.com to the straight-text style of www.bhg.com. Although you can choose to use graphics for your nav bars, consider the benefits of inserting text into a table with a background. First, it will be easier to change quickly, since you will only retype the text and make necessary link changes. If you use graphic navigation bars, however, every time the navigation for your site changes, you will need to create new graphics to match it.

Remember that once you choose a standard placement area for your nav bar, it should remain in that area so it can serve as a consistent reference point for users. When considering the final placement area for your nav bar, look through other top sites like those listed above. Most Web sites place their navigation bars on the left or top of the page. Newspapers have been taking advantage of this for a very long time; think about it: most people start reading at the top left of a page and work their way down. When determining where to place your navigation bars, consider the structure of your Web site. Most monitors display a greater width than height dimension, resulting in shorter screen views. If you place your navigation bar on the top of your page, ensure that it is small and does not eat up the valuable real estate known as "the first screen."

However, this should not keep you from considering the right location on the page for your nav bar placement. The bottom slot causes a few more problems, as information that does not appear "above the fold" (on the first screen) may be missed, especially on the front page. One definite advantage for right nav bars involves printing functionality. If you have not properly optimized your site for printers, a portion of the page's right side will be cut off when printing. Most people don't know they can avoid this process by selecting "landscape" as the printer type. If your navigation bar resides on the right side of your page and your page extends beyond the print margins, chances are your nav bar will not print. If you're printing out samples of page layouts for demonstration purposes, this may hinder your presentation. However, this can provide beneficial results—after all, how often do you think readers who print out pages for reading later will care about reading the nav bar? If

> **Nav bar (Navigation bar)—**
> *a site element that lists linked categories through which users can navigate a site*

you make your navigation bars distinctive, while still keeping them easy to use, you'll get the attention you seek and ensure that your users more successfully navigate throughout your site.

Working with Your Users: Feedback Tools

The most effective way to promote user interest in your site and increase usability involves adding some interactive elements that help you improve your site. Rather than limit your users to browsing through your Web site, encourage them to play an active role within your site. Establish an online poll once every six to twelve months that lets users tell you how well your site works and how it could be improved. Especially if you work with an intranet site (or provide incentives for external users), sit down with different users and watch how they navigate throughout your site. The feedback process enables you to understand the changes you may need to apply to your site for better usability.

4

Content to Keep 'Em Coming Back

Managing content for a Web site can provide great rewards and frustrations. Whether you work for a large company or own a small, home-based business, you may find yourself creating the first information seen by the public about the organization for which you work. Therefore, you must ensure that your site reflects timely, accurate information, effectively highlights the products and services your company offers, provides a straightforward, easily navigable structure, and that your content is free of typographical and grammatical errors.

Content management encompasses a variety of different roles in the conceptualization, creation, and launch of a Web site. First, content managers and producers should lead the programmers in the development of the basic site organization of the site, establishing usability standards, and maximizing meta tag information. During the creation process, content people should work on the development of the content with which the site will be filled, coordinate design elements with the graphic designers, and ensure the site organization continues to flow in the proper manner. Finally, content managers and producers need to review the site, checking for proper linking, work to finalize search engine submissions, and report usage statistics. All of those attributes and more require strong marketing, writing, editorial, customer relations, and project management skills.

> **Meta tag**—*a special HTML tag that provides information (the page creator, keywords, page description, etc.) about a Web page*

Top Tips for Content Management

1. Create an easily navigable site map.

2. Mind your p's and q's, and every other letter in the alphabet—spellcheck!

3. Keep it short, succinct—less is more, even in a Web environment.

4. Don't rely on text alone—energize your pages with graphics and lines.

5. Judiciously hyperlink to relevant information.

6. Create variety bullets.

7. Make content interesting and useful to the readers.

8. Avoid navigation challenges—help your visitors get where they want to go.

9. Keep content fresh—by providing updated, unbiased information on your Web site, you establish trust with your site visitors, and they will return to you as a respected resource.

10. Develop content for the Web—don't just cut and paste content created for print.

One of the most important aspects to creating great content for your Web site involves clear, effective writing skills. Be succinct and to the point. It's even more important to provide a quick read for Web site users, since most spend a very small amount of time on a Web site (often three links and 5-7 minutes!). Provide them with a quick and easy way to understand the major goals and highlights of your site. Use active voice. Help them "scan" the text for the most interesting and important material on your site. Don't lead your visitors in endless circles; guide them carefully to the content you wish them to read.

You may refer to other sites as a navigational aid, to support specific facts and information, or to enhance the content on your site. When you do, provide more than just links and the names of the other sites—offer a little background regarding WHY those sites should be important to your audience. For example, the bulleted references in this book should help you identify whether you would or would not want to visit the sites described. Also, identify and write for the audience that you wish to attract. Before you start creating new content, determine real demographics, such as age range, technical expertise, education level, language skills, geographic location, etc. You will create a richer, more rewarding content experience if you know for whom you are writing.

Additionally, no amount of careful preparation and glorious content will protect you if your site contains numerous errors. You must double and triple check your work—and we strongly advise you to rely on others for additional proofreading and editing. Nothing could prove quite as embarrassing as listing "pubic," as opposed to "public," but no spellchecker will catch that mistake. The references in Chapter 8, Troubleshooting Your Efforts, provide more information about content-checking skills that will help ensure your Web site meets content, as well as design, standards.

Managing Content

If you successfully establish a process through which you can manage the content on your site, you can directly affect the success of your site. Whether you need to acquire the approval of senior executives, clients for whom you are creating the site, or simply yourself, make sure the necessary people agree to the plan of action. Steps involved in content management include the following:

1. Determine the site organization.
2. Assess your content needs—what exists and what doesn't?
3. Identify primary audience and writing style requirements.
4. Leverage existing content, which can easily be adapted to your Web needs.
5. Research and produce new content.
6. Gain approvals for content, and if necessary, revise content to meet others' needs.
7. Push content to production server.
8. Optimize meta-tagging information for search engines.
9. Review content on production server.
10. Push content live or approve content for live push.

These simplified steps deliberately do not consider some of the barriers you may encounter in the review process. Some people will expect to see content only when it has reached its final stage and which you consider ready for production. Others will be satisfied with drafts; however, your best interests and sanity will be better served if you offer only final content to people. First, they will view your material with more professional respect if you have obviously prepared quality work. Second, by communicating up front that the material is in its final stage and needs only a cursory review, you maintain a higher level of control.

Keeping Control of the Review Process

Do your best to dissuade any changes after the information has been "Webified." Explain to reviewers that in order to save valuable time and resources, editorial decisions and changes should occur before you take the content to the programming stage. If you work as a consultant, identify and agree on two or three review rounds with your client, to restrict the number of miniscule changes. If you are fortunate to work with a team of talented individuals who program the pages and design the graphics, you need to help manage their resources by not overloading them with content change upon change. In some cases, of course, reviewers will need to review material in context with other elements, such as pull-out quotes and related graphics. That practice is fine, but attempt to set guidelines that minimize the number of picky stylistic changes at the end.

Small, personal sites may require fewer resources for content management and Web design. However, they need the same care and attention, just on a smaller scale. To help prevent unreviewed material from being posted to the public site, regardless of the size of your site or the organization it will represent, make sure you have a production (also known as staging or development) area and a live area. This will fortify your review cycle and encourage the buy-in from other participants, so it is important that you do not allow content to go live until it has successfully been through your review process.

Keeping Content Manageable: Tools That Help You Breathe

As your site grows, most likely your content will, as well. Therefore, you need to quickly determine how to keep your content manageable, before you lose control. Items to which you should pay special attention include older files, material that requires frequent updates, and material with links to external sites. Programs like GoLive and Dreamweaver will help you maintain some control over this information, by enabling you to quickly review the site structure, identify the dates on which the files were last changed, and correct broken links. Additionally, some of the steps listed in Chapter 8 will help you keep your content in decent shape, by staying on top of various testing processes. You could also use a calendar (electronic or traditional) to highlight the dates upon which specific content should change, or you could develop some other reminder system to identify on which dates certain files should expire.

Or, if you prefer not to play such an intensive part in working with the content of your site, you might consider moving to a database model for your content management. Content management via database frequently generates "dynamic content," enabling you to more efficiently track and manage content: who created it, who approved it, where it should post on your Web site, when it was created, and when it expires. You can either continue to serve as the central point through which content must flow, or decentralize content management.

By decentralizing content management, you identify others in the company to "own" specific areas of content. As content owners, they hold the responsibility for ensuring the material remains up-to-date, as well as follows style standards, as established by a style guide. This type of system works well beyond the corporate arena, as well. If, for example, you set up a "hobby" site for antique collectors, and wanted to encourage outsiders to provide information about special items, you could set up a database. By allowing people to submit information (either via restricted access or a special account), and encouraging them to follow certain style standards (you could prepare brief guidelines that must be read before supplying content), you create a content management network. Excellent examples of database use can be seen on ebay (www.ebay.com), Classifieds2000 (www.classifieds2000.com), and Inspired Living (www.inspiredinside.com).

How Content Management Programs Work Using a content management program, whether "home-grown" or purchased, will make your content easier to update, with reduced risks for technical problems, once the database has been configured to your needs. Either way will require some initial work from content and programming people, as well as some cost. Many businesses accomplish this objective using XML (Extensible Markup Language). Basically, set up a database like Microsoft Access or SQL Server. Make sure you clearly identify the information you wish to capture, such as author, expiration date, title, and fields for the actual content. After the users enter information into the database, the database will generate a text file.

Next, a special computer program that you have created or purchased will match up the data with a pre-created template. The result can produce either static HTML files or dynamic files. The first option will result in quicker load times, since the database will have already processed and saved the file. Although completely dynamic files result in slower load times, they do enable you to take more advantage of the personalization processes. Dynamic files require the database to search for and produce the file every time a user tries to open the link. This option produces a clear advantage in that your users will be restricted to using your site templates and you ensure that your site presents a standard look and feel.

Content Management Programs Pros and Cons Content management programs provide some distinct advantages, including a small learning curve for users, since information can frequently be submitted through any format, such as Word, Excel, etc. However, these and other content management tools offer more than just letting users enter information into a database so your Web pro-

> **Dynamic Files**—*files that perform a single action at the moment they are selected, rather than in advance, as with pre-coded HTML files*

grammers don't have to create each page. They automate your approval process, letting the content managers determine who must review the document before it can be pushed live—from legal to upper management, and so on. Along with this feature, content management programs provide "version control," a process which enables the content manager to track the users who have worked with the documents, as well as the changes that have occurred (think of the track changes function in Microsoft Word for a comparative example). If you work with multiple sites, such as intranet and Internet pages, you can also tell the content management tool to create versions for both sites. This results in a clear advantage—when the information needs to be updated, the person making the changes will only need to change one document, rather than two, three, or more!

Unfortunately, the powerful content management tools on the market, like TeamSite by Interwoven (www.interwoven.com), BroadVision (www.broadvision.com), and Vignette (www.vignette.com), all require significant resources—not only financially (six figures, in most cases), but also time and programming skills. Fortunately, some newer products are making their debuts and moving forward in the market; keep your eye on companies like Allaire, Percussion, and Rational for the product to meet your needs and budget. Some of the content management programs use proprietary coding programs, which will result in a steeper learning curve for your programmers. Also, content managers should not rely on these programs as the complete solution to content management, and should continue to conduct reviews of the live Web site to ensure content producers follow corporate styles; see the following section on style guides for more information.

Style Guides for Your Web Site

You can more easily reach consensus on content if you first provide style guides for people to follow, including yourself! They will help you quickly and easily define the manner through which your site presents content, including such crucial aspects as Web graphics, legal marks, and templates. Change your way of thinking about style guides, and view them as a repository of information about the styles appropriate to your site. Style guides can help internal and external users gain a better understanding of your stylistic requirements.

If you work for a large corporation, style guides will help protect your interests, as well as the consistency of your site, by providing guidelines available for internal and external use. If you work for a small business or maintain your own personal Web site, a style guide will help you organize the different templates, styles, and ideas for your Web site; if you follow it, you will give your site a more professional appearance. If you work with different clients as a Web consultant, access to the style guides of the companies you support will help you ensure that the Web sites follow their styles. Anyone who works with your Web site (or print materials) should be taught to rely on the style guide to ensure they are meeting your editorial, logo, and general image standards.

First, consider establishing an editorial board, which can review your suggested styles and offer new ones, essentially providing buy-in from the onset. In an ideal environment, this model will avoid future conflicts over stylistic issues. In a realistic situation, however, having initial buy-in from your editorial group and senior management will enable you to leverage those people in conflict situations and can reduce, not eliminate, style conflicts. Your editorial board can later serve to help evaluate content strategies, contribute future style guide insight, and mediate content issues. Whether you work for a large corporation with lots of resources, or cannot afford the luxury of establishing an editorial board because you work alone or for a small company, take a look at some of the style guides referred to below. They'll help you start your own style guide and provide some excellent insight on the different elements to consider. Upholding a style guide can represent a daunting task, but you must stand behind your style guide and enforce it—your site will be better for it.

Carnegie Mellon Web Publishing Style Guide
www.cmu.edu/home/styleguide
Fortunately for us, Carnegie Mellon includes this information-rich style guide on its public Web site. Standard elements include a basic welcome that introduce the purpose

and sections of the guide, and graphics, text, tables, and templates standards. However, don't miss the fantastic Writer's Style Guide, which covers everything from capitalization and "Using Tricky Words" to sensitivities and "Pet Peeves."

AT&T att.com Style Guide
www.att.com/style

AT&T provides a great site dedicated to agencies, people, and organizations who create Web sites for the att.com site. The standard creation guidelines include accessibility, organization, templates, and more. The "Publishing Process" provides guidelines for getting your information online, and help if you need it.

Your style guide may include a variety of elements from legal, editorial, programming, and design backgrounds, but it must seek to maintain two major goals. First, it should ensure that users follow certain standards so that even in a multiple content provider environment, the site looks consistent throughout. Second, your style guide should work as a dynamic document, changing as the goals of your company change, business products become renamed, templates are redesigned, etc. Additionally, it should remind people that content must remain fresh, timely, and relevant to its driving purpose. Depending on your needs, your style guide could incorporate some of the elements that follow.

Introduction: Corporate Philosophy, Boilerplates, and Templates

The first section should provide a basic introduction that establishes the reasoning behind your style guide, as well as stress the importance of using it. In this section, you can also provide an overview of the company's mission, value, and goals, as well as standard boilerplates that can be adapted for appropriate use. Also, provide some standard templates (PowerPoint,

Web, documentation, letters, etc.) which users can quickly access. If certain aspects of those templates cannot be changed, make sure you identify them up front.

Page Design and Stylistic Elements

First and foremost, the Web section of your style guide should describe the process through which people can submit content, as well as review cycles, graphic development, and a programming timeline. This section can also describe the types of elements you determined appropriate or inappropriate for your Web site. For example, if you and/or your editorial board have decided that animated images needlessly increase the load time of your site and therefore, decrease usability and should be not used, state so in your style guide. Additionally, provide your audience with the standard page dimensions and/or layout grids so they can effectively understand the size and placement with which they can work. If your logo must appear in the upper right corner of every page and no material should extend beyond the right margin it marks, and a top navigation bar must recur on every page, ensure that your users know this information.

You can also enhance the templates provided in the previous section, by presenting style guidelines for standard elements, such as Web-safe colors and sizes for text (headers, body text, links), backgrounds (Web-safe colors, images, etc.), frame acceptability (define when people can work with frames), tables (dimensions, use of nesting, coloring), and more. This section can establish that, in order to maximize the effectiveness of placement within internal and external search engines, <TITLE> tags must include descriptive wording, and provide guidelines for creating that wording. Use the Web section to describe coding standards, especially if you accept pages created and designed by people outside of your core Web group.

The Web section should also feature or link to a rogues' gallery, which clearly demonstrates how NOT to create Web pages for your site. You can create a rogues' gallery for each section of your site, or list all of the bad examples in one area, separated by category. Show examples that demon-

strate how bad coding can adverse-
ly affect the site, and discuss the
effects produced by different Web
design programs, as well as which
programs (GoLive, Dreamweaver,
FrontPage, etc.) do or do not create
acceptably coded pages. Firmly out-
law "coming soon/under construction pages." Often, the end user
who insists on including these will
provide the needed information
later, rather than sooner. This can
result in embarrassing consequences, and users will lose faith in
the timeliness of your site.

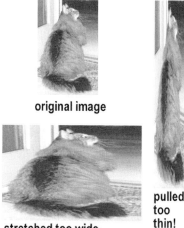

original image

pulled too thin!

stretched too wide--
makes a REALLY fat cat!

Examples of poorly scaled images

Web Graphics

Teaching users how to incorporate appropriate Web graphic styles can present some interesting and exasperating results. In most cases, few people
have some background working with graphics, or some resources (the
graphics chapter in this book or specialized training classes) through which
they can understand concepts like sizing and transparency. No matter how
carefully you prepare the graphic guidelines, you may end up using valuable time and resources trying to train the users how to properly prepare
graphics, or fixing mistakes. Not to discourage learning, but you may be
better off requiring users to submit original graphics, and letting the Web
team size them.

Regardless of whether you work with talented, design-oriented individuals or others, a graphics style guide will still assist you in setting some
ground rules. Sizing, both dimension measurement and file size, should
play one of the most important parts in your style guide. First, you need to
advise users of the largest and smallest images that can be used on a Web

site, arranged by category (header, specialized icon, front page picture, etc.). Hand-in-hand with this, explain the maximum allotted file size for each type of graphic, as well as the resolution necessary to create quality images. By showing examples of images that meet your requirements, you will help your users understand the standards a little more fully.

Stress that transparency must be in place for graphics that do not fit into a standard square, so the image will look the same, regardless of the end user's background settings. This also provides an excellent place to describe guidelines for video and animation—whether they loop once or several times, size requirements, etc. You can also direct users to more resources, including those described in Chapter 6, Your Best Image, to help them more fully understand the complexities behind Web graphic design. Finally, consider adding or linking to the rogues' gallery, which will show how inappropriate sizing can distort images, how non-transparent graphics display on a colored background, or how poorly programmed animated graphics and videos can look very amateurish.

Legal Guidelines

Legal guidelines should quickly and easily outline the proper use of company branding elements, such as logos, trademarked images, and service marks. This can overlap with the graphic guidelines when referring to logos, by advising users not to improperly size logos and providing examples of proper and improper sizing. Some people also prefer to list a maximum and minimum acceptable size for logos. In this section, insert downloadable versions of your logo, not only for electronic use, but also for print. The trademark and service mark sections should correspond to elements within the editorial style guide, showing proper naming and capitalization conventions, as well as providing (for internal use) a list of all intellectual property marks.

The legal section must also address issues relating to outside reproduction, first and foremost stressing the need to obtain permission prior to reproducing graphic, text, and other forms of intellectual property owned by the company or people by whom it was created, and the processes for receiving permission. It can also present the privacy and copyright statements that should appear at the bottom of every page, as well as legal documents that explain those statements in more detail. Finally, the legal guidelines should include a description of the process through which documents need to be approved. The description should carefully correspond with any guidelines presented by content people.

Editorial Guidelines

Editorial guidelines should focus on the preparation and application of text elements within your Web site. It can cover everything from tone and style of writing (don't get too "techy") to editorial processes, but should stress the mantra of fresh content representing an important aspect to any Web site. Some elements will correspond with the guidelines set in the legal section, including proper references to your or your company's intellectual property, as well as proper use of others' intellectual property (covered later in this chapter). And, although it might seem like a simple concept that everyone should know, clearly state that defamatory, obscene, and other inappropriate material should never be introduced into the site. This statement should prove especially helpful when relating to sites that accept outside content submissions, yet do not maintain constant editorial control over that content.

You can also maximize the editorial style guide by listing boilerplates appropriate for a company, products, and services, as well as identifying (and updating!) standard facts about which everyone should be aware. Identifying senior executives and their biographical information, or linking to that information on an intranet or Internet site, will help people quickly and easily access the latest information for those frequently referenced individuals. While you're explaining those standards, provide some guidance on how to properly reference people by their titles; the standard method lists titles in lowercase but capitalizes departments. Whichever you chose, make sure all content remains consistent in this matter.

You can also incorporate a grammatical/proofreading reference section to help users identify and navigate trouble spots. Make sure that you provide clear examples for each, in the form of "before" and "after" examples. Essential elements of this style guide area include capitalization, punctuation, and possessives. Advise your content contributors that passive voice creates a weaker style of writing and show them how to convert this information into active voice. In the process, you will help create better writers out of your content providers. Consider a "word guide" that identifies trouble words, such as abbreviations, dates, numbers, places, product names, frequently misspelled words, and other words with which you notice persistent problems.

Acquiring Content

You can acquire content for your site through a variety of internal and external resources. If you work with individuals within your company or contractors who produce work for hire, you will encounter significantly fewer legal concerns, as the information will be specifically prepared for your site. Additionally, if you are a consultant working on a Web site, stress from the onset that you need access to the knowledge bases, too—not only those people who will create content, but also the resources (people and publications) that can provide the information you need. When working with those resources, your greatest challenge will be assuring that people provide content on time, and that the quality of the content matches your style guidelines. Provide easy access to your style guide, conduct editorial planning meetings, and ensure that people meet mutually established deadlines.

On the other hand, securing content produced outside of your company's employees and network of consultants does not need to present a major headache. If you discover

> **Passive voice**—*writing in which weak, or "helping," verbs appear in the sentence, creating a less direct result. Passive: Our decision was to stay overnight. Active: We decided to stay overnight*

information on the Web that you would like to reproduce on your site, you may surprise yourself with the ease by which you can gain permission to use the content. In many cases, a simple query letter to the Webmaster of the site will result in the point of contact needed. In your request, identify your company and/or Web site, provide some context as to how the information will be used, the length of time intended, and agreement not to reproduce the content beyond the indicated intent. Also, inquire about the procedures they require in order to reproduce their content. Approach them in a self-assured, professional manner, with the assumption they will give you the permission you seek, and you will improve your chances of success. In some cases, they may request that you link to their site or include a small reference indicating the source of the information. If you agree to link to their site, keep in mind that you run the chance that visitors may consider the linked site more interesting and not return to yours!

In addition, special content providers exist purely to create content for Web sites to reproduce. Many of these groups charge significant fees for the content, which can include everything from horoscopes and weather to stock quotes and competitive intelligence. Take the time to interview and compare them to determine the sources from which they derive their information, as well as the types of information they can provide, and the flexibility of content determination. Other considerations include the frequency with which the information can be updated, ranging from media feeds which get updates as companies file their stories, to daily or weekly content; the cost; and the ease with which this content can be applied to your Web site. Below are brief reviews of some content companies that you may wish to approach:

Yellowbrix
www.yellowbrix.com
Yellowbrix will either host or partially host your service through a variety of delivery methods, including XML, HTML, and JavaScript. You can request that they personalize content and deliver it through Web, email, or wireless

channels, which increases your audience range. Topic categories include news, company intelligence, stock quotes and charts, weather, horoscopes, photos, and more.

iSyndicate
www.isyndicate.com

With over 4,700 resources, iSyndicate provides a robust repository of information from which you can draw interesting content to bolster your Web site. Their interactive site lets you quickly and easily determine content relative to your needs, and have that content delivered to you in a variety of ways. Content areas contain information about health and fitness, weather, international, games, comics, weather, and much more.

Subportal
www.subportal.com

Through this handy Web site, you can add free content, while keeping your Web site's look and feel. Subportal also provides you with the opportunity to earn money, retaining 25% of revenue generated by banners from paid advertisers. Content topics include news stories, recipes, and horoscopes.

Moreover
www.moreover.com

Moreover lets you add dynamic news headlines to your Web site in only five minutes. By selecting from over 1,000 categories or building a new category from your keywords, you can show your users constantly updating news headlines. The process is achieved by designing your Web layout and integrating provided HTML code into your Web site.

Common Site Elements

If you follow this book chapter by chapter, your site map, directory structure, and content may already be determined. If not, here's an opportunity for you to consider a variety of fun content elements that will enhance your site visitors' experience. Ideally, you want to provide content that supports your Web goals and audience interests, but you may want to explore beyond your initial site conceptualization, especially once you have launched your initial Web site attempt. Since sites always need to vary in information if you want to convince users to return, following are both standard and unique ideas that will excite your visitors and keep them coming back to learn more.

Internet Sites

Whether for personal or professional use, Internet sites can provide users with a variety of content that can impress and inform them. The most important challenge to determining content for your Internet site involves a careful balance between fun and necessity. If you work for a corporation, make your site more than a marketing or sales machine that serves as a catalog of your products and services—reach out to other audiences, such as investors, media, potential employees, and the general public, by providing educational, informative material about your company. If you maintain a small site through which you provide information about your children and their amazing achievements, consider adding information about books that helped you achieve your parenting skills, or an interactive game that lets others determine their level of parenting expertise.

About Us Establish the who, what, where, when, why, and how of your site through this informative section. Include bios of the people relevant to your site mission. For example, on a genealogy site, you might explain your background in genealogy, how your interest started, or a brief background of your personal genealogy. On the other hand, a professional site would list the executives in your company, including their present work responsibilities, previous work experience, education, board and professional organization membership, philanthropic outreach, awards, and significant achievements. Also, explain the reason for your site and/or company, such as its history, mission, and goals.

Employment If your site represents a company, especially one with an interest in hiring qualified individuals, consider adding an employment section. This section can save the company recruiting and job advertising fees, by helping your Human Resources (HR) department hire quality candidates. Definitely include available jobs (ensure that recruiters can easily access and update this section), as well as a way for candidates to apply for the jobs online. You could also ask screening questions, such as years of experience and technical experience, to help your HR personnel pre-determine qualified candidates. Additionally, this section of your Web site should help candidates learn more about your company, by providing information about benefits, corporate culture, etc.

Press Room Journalists, the main audience for this area, usually run on tight deadlines and lower-end computer resources, so this area should provide quick and easy access to relevant information, such as press releases, fact sheets, metrics, FAQs, and links to the about us and investor relations areas. PDFs and graphs require longer load times, so use them judiciously. Press releases, the main focus of this area, might highlight new products or services, new executive hires, elections to industry boards, big achievements, awards, or partnerships. Press releases must include contact information, or it should be easily located on the site to help the media quickly access resources to check facts.

Offer reporters the ability to browse releases by topic area, but definitely list them by dates, as well. When using dates, use reverse chronological order, and remember that the world does not identify dates in the same manner. For example, most people in the U.S. view 01-03-01 as January 1, 2001, whereas to a good portion of the rest of the world, it represents March 1, 2001. To avoid problems such as this (and ensure that both Americans and the rest of the world understand the date you are trying to identify), try a different format, such as Jan. 3, 2001. This will help the media quickly and easily determine the timeliness of the release.

Contact Us Users will invariably want to contact you, whether to speak to customer service, inquire about a job, or to ask a variety of questions you thought your Web site clearly answered. Make your contact information easy to locate on your Web site; definitely include it on your front page and possibly include a link on your global nav bar. You should include telephone numbers, fax numbers, and email addresses for a wide variety of contacts, including HR, investor relations, media relations…and the list goes on. Additionally, to improve your customers' experience, build out your customer service number list to identify focuses, if possible, such as technical support, new account set-up, and product returns.

Example of Contact Form page

To protect the privacy of your employees, however, avoid listing specific contact names on your site. For your investor contacts, you might list a general email address, such as investor@topsitesforyou.com, which can be accessed by a specific employee or group of employees who will ensure the message gets addressed by the correct person. A good contact management system will quickly enable people to identify standard contacts, while decreasing the chances that contact attempts will be directed to the wrong person or place your employees in awkward situations.

Events Maybe your company appears at a lot of trade shows, attends career fairs, or conducts extensive training sessions. Perhaps experts in your company speak in a variety of forums, such as association meetings or college classes. Or, maybe you want more visitors to your site to see your speaking expertise. Use your Web site to highlight your events, creating a calendar that lists information and appearances by date, type of event, and/or topics.

Products and Services One of the most important areas for any Web site that features items or services available for sale, your products and services section needs to provide more than the names and costs of your products and services. Enhance this section by including fact sheets, product specifications, manuals, and multimedia demos. You also need to provide contact information in this area, including relevant numbers for sales people and customer service. Speaking of customer service information, don't forget to include downloadable patches and support information, which could appear as searchable areas of product support content or FAQs. Finally, make sure that visitors can easily obtain your products, by offering secure, real-time purchase and registration services.

Customer-Related Information If you believe in your products and your company, take that belief one extra step and encourage communication between your customers. Highlight success stories and testimonials, and create a special area through which customers can add similar information. Information for these areas is best monitored before it is posted to the site for the world to read. These pieces of information can reassure new and potential customers that someone other than you recommends your products, services, and expertise. To more effectively capture such information, as well as provide customers with a way to notify you when something does not work well, set up customer feedback forms that they can fill in and have automatically sent to you. Make sure your system sends them an automatic response to their feedback, and that whomever receives the action item responds and resolves the problems quickly.

If you're completely ready to go out on a limb for your company, start mailing lists or bulletin boards through which your customers can exchange information. You can easily monitor these tools, ensuring that only appropriate information gets posted to the site. Post guidelines for content submission (profanity, information mining, and harassment should not be allowed!) and if you decide to "moderate" such areas, ensure that you make that policy clearly available, otherwise you will receive complaints.

Intranet

When creating content for an intranet, follow the same basic procedures as those used for Internet content. Of course, the biggest advantage provided by an intranet over the Internet involves the level of privacy and control over the content within the network. You can publish competitive information, corporate documents, and special product and personnel information not appropriate for external users. Your intranet can also serve as a resource for customer service and management, increasing the communications between the employees within your company to reach for a higher customer focus. Also in the interest of peer communications, set up resources through which employees can share information, such as leads, presentations, customer and product information, and "war stories." The following resources provide some extra background for creating and maintaining a successful intranet.

Intranet Design Magazine
http://idm.internet.com
This helpful site borders on the edge of overwhelming you with resources, primarily articles, on everything from HTML coding to wireless intranet development. If the search engine and well-organized categories don't fulfill your information search, check out the intranet FAQ, glossary, and discussion board—or turn to other listed resources, such as the events calendar and intranet book listing.

CIO's Intranet/Extranet Research Center
www.cio.com/forums/intranet
CIO presents a great repository of case studies, relevant articles, and research sources. If you still want more help with your intranet preparation, turn to the online seminars, CIO radio, discussion forum, or the variety of other useful links to other Web sites.

One of the most enjoyable aspects of intranet Web design involves standardization. As opposed to working with an Internet site, where you must limit your creativity to designs and applications that every browser can see, with an intranet, you can usually create pages for one standard platform. On the other hand, your content can vary SLIGHTLY more than the content exhibited on the Internet. Since the audience should have a less formal bent, you can more easily trust different content providers to provide and maintain content that follows corporate standards and move to a more decentralized content management practice.

If you can, determine a way through which users can connect to your company intranet remotely, so they can access the useful information you provide any time they need it. Especially if you can provide this valuable access, ensure that your company's privacy policy/terms and conditions are listed at the bottom of every page. An intranet can reduce costs of printing and distribution of forms, presentations, and policy documents, as well as personnel time—both for printing/collating and for preparing special formats. Expand upon this useful tool, consistently improving and updating the information so that your employees regard it as an invaluable resource. Special areas through which you can enhance your intranet include employee suggestion forms, training resources, an electronic paycheck utility, software manuals, vendor information, maps, a style guide, and more. Additional special areas through which you can increase information include:

Employee Directory Any way you create it, an employee directory will prove one of the most frequently used tools on your intranet. Your employee directory can be as simple as an Excel file or HTML table that someone creates and owns. However, if your company has the resources to create or purchase an online application that enables people to make changes to a dynamic database, this tool will quickly make up its cost and start saving the company money. You will be able to decentralize the ownership for individual employee records, though the overall directory should still have a central point of contact for employee questions, new-hire record entry, and special changes needed to the directory.

Although you can identify a primary person responsible for entering new employees into your employee directory system, employees should own the responsibility for updating their own records. If you empower your employees in this manner, take some basic precautions to ensure that each person can only edit his or her own data. One possible way to achieve this goal: Prior to each change made to the database going into effect, an automated email will be sent to the employee's company email account; in order for the change to occur, the employee must respond to the email.

Standard fields that should be included in every directory might seem obvious: first name, last name, phone number, email address, and fax number. However, think beyond the obvious and add supervisor and department data. Suddenly, you enable employees to contact supervisors if an issue needs escalation or a personnel conflict occurs, or, conversely, to contact an employee's manager to recognize special achievements or appreciation for a job well done. By expanding department data to include job specialties, you create an experts' database that can help other employees quickly and easily find the answers to customer inquiries. If you expand your database even further to include skill sets, you completely utilize your employees by identifying previously unknown language skills and special interests—no matter how unusual or unimportant those abilities may seem, they can one day prove helpful. For example, you might need someone proficient in German to interpret an email to the chairperson of your board (it's happened!).

Help Desk Through a help desk, your employees can quickly and easily reach a variety of support areas within the company. This area should include self-help capabilities, such as access to special forms (HR, payroll, equipment requisitions, etc.). Also, consider adding FAQs that identify the intranet resources that will quickly answer questions regarding standard company policies and procedures. Finally, intranet help desks should include interactive forms through which users can submit requests for HR, information technology, or facilities assistance that cannot be resolved via the resources on the intranet.

Human Resources (HR) One of the focal points of an intranet, the HR section helps employees maintain their jobs and related benefits. A well-designed HR section should include detailed benefits information, such as plan enrollment information and online registration capabilities, types of benefits, and benefits forms. Additionally, the "official" benefits handbook can easily be converted for Web use, saving printing and mailing costs in the process, and should include plan prospectuses and details about related topics. Other important areas in the main HR section include a listing of employee events and due dates, the holiday schedule, and HR staff and external plan management contacts.

The employment area of the HR section must feature an internal listing of available jobs, or a way to easily access the external employment site. It should also detail your referral program, if you have one, helping employees efficiently submit candidates to HR and receive an auto-response recognizing their contribution. This area can also prove especially useful during your employee review period, by providing review forms and helpful tips to help employees effectively prepare for this process.

Internal Newsletter An internal newsletter lets companies highlight special news relevant to the needs and interests of employees. Make sure that senior management supports the newsletter and uses it as its main resource for distributing company information. By publishing this information on a regular basis, you can draw attention to important company information, such as events, new products, and public attention received by the company. Other interesting features to include in the company newsletter focus on new employees, special company Web sites (internal and external), or featured "employees of the week." You can also encourage interactivity with the newsletter, by offering games or trivia contests.

Sales and Marketing Information The sales and marketing area on a corporate intranet can provide valuable resources through which salespeople and marketing departments can improve their understanding of the customers and industries with which they work. Standard information for this area should include competitive industry news and information, as well as relevant market research and links to your own company sites (especial-ly press releases and your main external site). This information can be expanded to feature best practices and lessons learned. Finally, consider expanding this area to offer sales and marketing leads and customer profiles. Sales and marketing representatives from different areas of the company can leverage each other's contacts to improve company revenue.

Customer Service Area By providing your customer service represen-tatives with a dedicated section of the intranet, you can help them decrease response time and improve their knowledge. An easily accessed inventory and supplies for an organization that sells products will help customer service quickly determine the availability of specific items. By providing an online customer database accessible for anyone who directly interacts with cus-tomers, you enable them to immediately determine the location and pur-chase record of each customer. They also can review the history of their rela-tionship with the company and identify specific "special" customers who may display difficult behaviors or a great purchase record.

Interactive Content

By creating interactive content, you can engage and capture the interest of your site visitors. Rather than merely giving them screens of information through which they can read or ignore as they navigate around your site, consider providing them with the chance to participate. Standard interac-tive content tools include printing, bookmarking, searching, and filling out online order forms. You can also add interactive tools to your pages of content, enabling users to automatically email the page to a friend or col-league, download the content as a Word document, or leave feedback about the particular Web page. Additional interactive tools let users sub-scribe to receive relevant newsletters and updates by email, guest books, specialized registration or feedback forms, surveys, bulletin boards, polls, and more. Take a few minutes to consider some of the easy tools you can quickly offer your users:

Surveys/Polls

Surveys can be used to gather information about the people who visit your site, and can result in marketing lead generation, improved demographic data on your visitors, new ideas for your site, and many other possibilities. Keep them as short as possible; ask only the most important questions to make the process quick and painless for respondents. If a survey is longer than a standard screen's length, disclose up front the amount of time that should be necessary to complete the survey. Otherwise, you may frustrate users who thought they could complete your survey in a few minutes before their lunch date, which could result in partially completed or deleted surveys.

Surveys providing multiple choice answers give you more feedback than yes/no answers. They also are quicker and easier for users than those that require individual, unique answers, as in a fill-in-the-blank format. However, for every multiple choice that lists "other" as a choice, make sure you leave a blank so that users can specify an answer not identified in your list. Consider using an outside company with survey experience; they have the knowledge to pull together good behavioral questions that will help you get the results you seek. Most will also interpret the answers for you, which will quicken your data interpretation process.

Infopoll
www.infopoll.com

This interesting site guides you through surveys and polls as information-gathering tools, and includes samples, from customer and market research surveys to political opinion polls and feedback forms. The site also highlights products to help you design, simulate, deploy, manage, and analyze your surveys. Additionally, Infopoll offers information about more resources, including survey-related books, articles, and market research Web links.

WebSurveyor

www.websurveyor.com

Start with a free account to gather some great ideas for your survey needs, including a great how-to section that walks you through the survey preparation and analysis processes. This site also provides some interesting white papers, articles, and suggested books to help you on your way to survey perfection. Finally, check out the prepared surveys, which will give you a better understanding of this site's capabilities.

Guest Books and Registration Pages

Take care with the creation of your guest books and registration forms. First, determine what content you really hope to capture from whomever is filling out the form. If your goal is a simple guest book that identifies people who have visited your site, help them protect their privacy by capturing their first name and general location. If you want to learn more information, ask how they heard about your site. On the other hand, consider expanding your guest book to capture more information about your visitors, including their interests, their Internet experience level, and even their favorite colors. That information, combined with feedback solicitation about your Web site, can help you improve the content on your Web site.

If you work for a larger company that offers training, informative publications, or subscription newsletters, follow the same rules of thumb for registration forms as those suggested for guest books. Without overwhelming your users with questions, capture their

Example of Guest Book page

name, company, and occupation. If you decide to request information about household income, company revenues, or other financial questions, you should make those questions optional. Many people deem financial matters private and will not complete out a form that requires them to provide that information. By offering the opportunity to "opt out" of sensitive questions, you run greater chances of success in any registration forms being submitted.

Order Forms

As described in the guest book/registration area above, order forms should be quick and easy for site visitors to complete. Determine the most important information, such as name, phone number, email address, and mailing address. Make sure that every order placed on your site generates an automatic response that verifies the order has been placed and provides a confirmation number—preferably through email, to most effectively reassure your customer.

If you decide to use this information to generate leads for more sales or to sell the gathered information to marketers outside of your company, don't be secretive. Inform your customers of your privacy statement, and make sure you include a method for users to declare that they do not want their information to be used for any purposes beyond the specific order. Additionally, if you or someone else have not already done so, identify your site's security, so that people will feel more comfortable providing payment information.

5

Code Challenges

Behind every good Web page hides some hard-working code called HyperText Markup Language (HTML). HTML code tells your Internet browser how to display all the "pieces" of your Web page, such as images and text; for example, how big to make them, where to put them, and what color. Through some time and effort on your part, you can learn how to master HTML and create great Web sites.

And, yes, you can use a friendly software package or an even friendlier Web site to do your coding for you, but it's always good to understand how to write some basic HTML. It's not as difficult as you might think, and with only a few basic "codes" or "abbreviations" under your belt, you can create a Web page from scratch. You'll need four essential tools to create HTML code and make a Web page "live":

- ◆ A text editor—to create and edit your Web pages

- ◆ A Web browser—to preview your Web page before it goes "live" online

- ◆ A modem and Internet connection—for eventually posting it up on the World Wide Web

- ◆ A place to host your Web pages—whether your own server or an outside hosting service

HyperText—*text that is linked to other text or documents so that when you click on it, you're transported to another location*

Since it's going to take a little bit of effort, thought, and time, you may well be asking, "What kinds of nifty stuff can HTML really do for me and my Web site(s)?" For starters, it can jazz up your text by creating different font sizes and styles, including bolding and italicizing. HTML will enable you to enhance your pages with colors, lines, and graphics. You can also excite and intrigue your visitors through the addition of animation, sound, and video files, or take them to other great Web sites through hyperlinks. You'll be amazed at the effects a simple knowledge of HTML can produce. Computer training courses will give you a more thorough understanding of coding Web pages, as will the following books:

- *HTML 4 for Dummies*, by Ed Tittel, Natanya Pitts, Chelsea Valentine

- *HTML 4 for the World Wide Web Visual Quickstart Guide*, by Elizabeth Castro

- *HTML & XHTML: The Definitive Guide*, by Chuck Musciano, Bill Kennedy

On the other hand, if you want a true Web experience when creating your Web site, turn to the Web for excellent examples of HTML use. Find pages you like and view the page source to see how the experts work. You can also browse through the treasure trove of Web sites that offer assistance with tags and how to use them. Here are a few where you may find useful references on HTML:

The Beginner's Guide to HTML
www.ncsa.uiuc.edu/General/Internet/WWW/HTMLPrimer.html
Use this helpful guide as your starting point for learning about HTML. Published by National Center for Supercomputing Applications (NCSA), this resource provides one of the most comprehensive online HTML tutorials. Although NCSA has ceased updating the guide, it still remains a vital historical and educational document regarding HTML use.

The Bare Bones Guide to HTML
http://werbach.com/barebones
The Bare Bones Guide lists every official HTML tag in common use. You can view or download the "Bare Bones Guide

to HTML" in several different formats and 21 languages. Version 4.0 of this comprehensive guide conforms to the HTML 4.0 specification of the World Wide Web Consortium (W3C). The site's "WWW Help Page" has many excellent resources and information on other Web design topics.

HTML Goodies
http://htmlgoodies.earthweb.com

Check out these great online HTML tutorials for both the absolute beginner and the more advanced HTML user. The HTML goodies are divided into groups called Primers, Tutorials, and Beyond HTML; the latter group contains advanced information such as Webmaster tips, JavaScripting, ASP, XML, and much more. The seven-day plan of learning essential HTML through a series of seven HTML Primers is easy to follow, painless, and informative. As a bonus, the site is well laid-out, professional, and easy on the eye.

HTML Cheatsheet
www.webmonkey.com/reference/html_cheatsheet

Learn all of the essential tags, listed in a clean and simple layout on this HTML cheatsheet. Tags are neatly grouped into sections such as Basic tags, Formatting tags, Table tags and attributes, Forms, and so on. Additional quick reference links connect you to an excellent glossary, a JavaScript code library, a stylesheet guide, browser library, and much more.

Free Site Templates
www.freesitetemplates.com

If you don't feel that you're quite ready to create your own HTML code, this site has a great selection of HTML templates that you can download for free. You can display them first to see which one you like.

Tags—They're It!

So, how **do** you give the browser instructions on how to display your Web page? Use "tags"—the codes that instruct the browser to display your actual text and images in a specific way. Once you know a few basic tags, you have a lot of power at your fingertips and will enjoy trying out different ways to show off your skills through a high quality Web page.

A tag is a code with one or more letters; it can appear as an abbreviation or a whole word enclosed within less-than and greater-than symbols (< >). Tags usually (but not always) come in pairs and surround your text in the following format: <TAG>Text to be affected by the tags</TAG>. Notice that the second tag has a forward slash "/" (never use a backward slash "\") in front of it, which identifies the tag as a closing tag. One of the most commonly made programming mistakes occurs when the programmer forgets to insert the closing tag. Forgetting this tag will "confuse" the Web browser and it will probably not display that piece of text in the way you want or expect. Here's an example of using a tag to make a piece of text appear in bold type: <B>100 Ways to Make Money</B>. If you forget to use the </B> tag at the end of the text you wanted to bold, the browser will be confused, and will bold the rest of the text in the document until it reaches the final </HTML>tag.

It doesn't matter whether you use all uppercase letters or lowercase for your tags. Either will work, although using all uppercase letters does help to make the tags stand out from your text. You'll more easily be able to spot uppercase tags when you're looking through your HTML code.

> **Element**—*the basic instructions, or "tags," that make up HTML source code*

Basic Structural Tags

+ <HTML></HTML> Start and end your code with these tags—they enclose the entire HTML document and tell the browser that this is a file that contains HTML.

- HEAD></HEAD> These tags define the header (the document can essentially be divided into 2 sections: a header and a body), into which you will place tags like <TITLE>.

- <TITLE></TITLE> Insert your title in between these tags, and always place them between the opening and closing <HEAD> tags. Browsers will display the resulting title at the top of the browser window, rather than within the physical Web page. Search engines will often use the content of the <TITLE> element to list your Web site.

- <BODY></BODY> Insert all the other text and graphics in between these tags. Remember that the main part of your Web page is going to consist of the text and graphics that you put in between the <BODY></BODY> tags.

- <H1></H1> These tags display the text between them as a major heading. (Note: this is not the same as the TITLE, which is displayed in the browser window.)

Here's a skeleton file that you can use to start off your Web site. It contains just four basic tags and some sample text that build the essential structure of the Web page—rather like laying the foundations of a house.

The Skeleton File

```
<HTML>
<HEAD>
<TITLE>Insert Your Page Title
Here</TITLE>
</HEAD>
<BODY>
<H1>Insert the Main Heading of
Your Page Here (Optional)</H1>
type in some text
insert some pictures
</BODY>
</HTML>
```

Example of Skeleton File

Tips:

1. It makes no difference if you use uppercase or lowercase for the tags, but it's much easier to read if you use uppercase.
2. Use each of the basic six structural opening and closing tags only once:
   <HTML><HEAD></HEAD><BODY></BODY></HTML>

Basic Formatting Tags

Make your text stand out from the crowd by using tags that format your text in creative ways. Give your Web page a professional and finished look through the creative use of fonts, which quickly and easily let you change the appearance of the characters in your text. Experiment with such fonts as Times New Roman, Verdana, Arial, and others to decide what looks best.

Tag Pair (Begin Tag, End Tag)	Text between these tags is displayed as:
<P></P>	New Paragraph (closing tag is optional)
<I></I>	Italic
<B></B>	Bold
<U></U>	Underline (use sparingly, as this may confuse users with hyperlinked text, which frequently appears underlined)
<TT></TT>	Monospace (looks as though done on a typewriter)
<FONT></FONT>	Changes the font type—you can select from a variety of different styles, sizes, and colors (see below)

Tag with Attribute	What It Does
<FONT SIZE = "n"	Sets the size of the font (the default font size is 3)
<FONT COLOR = "color">	Sets the color of the font
<FONT FACE = "Arial">	Sets the font face (the appearance of the lettering)—in this case, to **Arial**. Other common font faces used on the Web include **Helvetica**, Times New Roman, and `Courier`.

Other Formatting Bits and Pieces

Tag	Name	What It Does
<CENTER></CENTER>	Center tags	Centers everything inside the start and end tags
 	Line Break tag	Browser displays line break after this tag
<HR>	Horizontal rule tag	Displays a horizontal line

Sometimes, you need to add just a little more information to your tags to give some very specific information to the browser. For example, you may want to place a horizontal line on your page, but what if you don't want to center it? Perhaps you want to line it up with the left margin? You can define this process by including an extra piece of information inside the tag's brackets (<>) called an "attribute." You should follow the good practice of surrounding your attributes with quotes; in some instances, browsers will not catch the information unless you use quotes (for example, where an attribute consists of more than one word). In the preceding example, we showed some of the attributes you can use with the <FONT> tag. Here are some examples of horizontal line tags used with attributes.

Tag with Attribute	What It Does
<HR ALIGN="LEFT">	Aligns the horizontal line with the left margin. You can also indicate right or center for alignment.
<HR NOSHADE>	Displays the line as a solid line (instead of appearing as etched into the screen)
<HR SIZE="n">	Sets the thickness of the line to n pixels (the default is 1 pixel)
<HR WIDTH="x">	Sets the width of the line to x pixels
<HR WIDTH="x%">	Sets the width of the line to x percent of the screen

Headings

There are six basic levels of heading, which decrease in size as they increase in numerical value. For example, an **H1** heading is larger than an **H2** heading. Another interesting tidbit: Headings include automatic spacing after them, so you do not usually need line breaks or paragraph marks to differentiate them from regular body text. To apply headings to text, follow this formula:

<H1>Level one header text</H1>

You can select from header levels one through six. Make sure you surround your header with both the start tag and ending tag, or you may encounter different effects than you expect. You can also add attributes to header tags,

header 1

header 2

header 3

header 4

header 5

header 6

Examples of header styles

similar to the <HR> tag, to modify its position on the Web page. For example, <H1 ALIGN="center">Level one header text centered</H1>.

You Are What You Link

So, now you know the basic tags that form the backbone of a Web site. However, a one-page site that leads nowhere else will most likely receive little attention from the audience you want to target. Hyperlinks (links) add "depth" to a Web site by enabling the user to navigate within a page, to connect to other pages within that Web site, or to visit external sites—all through a little bit of coding. As covered in the following subsections, internal and external links achieve different results; you should carefully plan your site to ensure that you achieve an easily navigable site. See Chapter 3, Good Design Basics, for guidance on creating effective site maps. About.com (www.about.com) and Happy Puppy (www.happypuppy.com) show broad diversity and good handling of internal and external links.

Internal Linking

Internal links (often called "bookmarks," though not to be confused with the Netscape browser's usage of the term) result in a link on a page that connects to a "target" section within that page. This type of linking proves especially helpful when you work with a very long page that needs to remain a single document. One reason for such a decision is related to printing—by containing all of the information within one page, the Web designer does not need to design a new page that can be printed out as one document. On the other hand, perhaps you're working on an online newsletter, and for site management and archival purposes, one page meets your needs. By creating an "index of links," strategically placed somewhere in the document, the user quickly and easily navigates to the section he/she wishes to review, rather than scrolling through seemingly endless pages of text.

Internal links require the creation of two codes in different areas of your document. The first code defines the location of the internal link—a bookmark of sorts, as it holds the place to which you will return. For this coding, use the **<A NAME="text"></A>** tags; they should be placed in front of the section to which the link will jump. When using this to jump people to a specific section of a document, you might want to include the title of that section, if relevant. To create the actual link, you'll insert, as the second code, <A HREF="#text">Sample</A> around the text that will serve as the hyperlink. Therefore, when the user clicks on the hyperlinked text, "Sample," the browser will interpret the link and take the user to the section bookmarked as "text" later in the document. You must use good organizational skills for this process, and ensure that each unique "#text" link on a page corresponds with the correct "text" bookmark.

External Linking

On the other hand, perhaps you would like to expand your site beyond one page—always a good idea in today's competitive Web design world. Obviously, you can consider two types of sites to which you wish to link: those which fit within the subdirectories of your overall Web site, or those completely external to the site you have created. While working within your family of Web pages, you'll encounter few outside factors that may affect the operation of your site; your greatest concern will be your troubleshooting abilities.

Link to other pages within your Web site by placing the text you want to use as the hyperlink between the front tag and the end tag <A HREF="directory/pagename.html">sample text</A>. If you are working within the same directory, you need not indicate the directory within your front tag. This link is referred to as a "relative link," since it works through the relationship between the source and link files. Using a forward slash by itself, <A HREF="/"></A>, will link to the "root" of your Web site. If you choose to link to a file outside of your Web site, you will need to create an absolute link, or a link to the complete (absolute) URL for the particular site to which you will link, with the start/end codes: <A HREF="http://www.name.com/dir/page.html"></A>. Some designers choose to use absolute links for sites within their directory structure, which we do not advise. Absolute links point to a very specific location, and if your site structure changes, you will likely encounter a lot more administrative work if you use absolute links.

Working with pages outside the realm of your Web site may result in a variety of new challenges—from broken links to legal implications. Wherever possible, try to choose high-level links that do not travel down too many subdirectory paths. As other site managers change the structure and file naming conventions for their sites, any links you have to those changed pages will fail to resolve, or will "break." For this reason, you need to ensure that you frequently check your site for broken external links.

Of course, the most important (and sometimes, fun) decision involves choosing great sites to which your site will link. Try some of the search engines listed in Chapter 1, Getting Started; they offer great resources that may prove relevant to your site. If you work for an Internet company, for example, you may want to link to legislation or news articles relevant to your interests. Keep in mind, however, that every external link you list sends a user away from your site. Once on the external site, the user may "forget" to come back to your site, and you have taken the chance of lost revenue or interest. One workaround for this (which should be used sparingly to lower user frustration levels) results in an additional browser window being launched whenever someone clicks on an outside link. To code this, just add a "blank target" to your link: <A HREF ="http://www.name.com/dir/page.html" target="_blank">. You can also apply this linking principle to pages internal to your Web site.

Tag Pair (Begin Tag, End Tag)	Resulting Action
<A NAME="text"></A>	Anchor tag/"bookmark" for link
<A HREF="#text"></A>	Link takes user to pre-defined "anchor" within file
<A HREF="dir/page.html"></A>	Relative link transports user to another page within your Web site
<A HREF="http://www. name.com/dir/page.html"></A>	Absolute link sends user to another page, either within your Web site or on another Web site
<A HREF="..." target="_blank"> </A>	Target opens a new page when user clicks on link

Did you know that lawsuits have been filed by people and companies unhappy with external sites that link to them? It may sound funny—after all, why would anyone mind having another site link to his or her site? Unfortunately, the complications and tensions regarding this issue involve many different facets of which you should be aware. One of the biggest complaints arises from sites with conflicting interests linking to another site—for example, a pro-drug site linking to an anti-drug site's list of drugs and their effects might raise some eyebrows on the side of the anti-drug site. In many situations, legal actions are not taken; however, if you really want to play it safe, request permission before linking to a site.

Lists and Bullets—How They Make Your Information Stand Out

There will be times when you want your text to really stand out—whether as bullets or a numbered list, as seen in traditional business documents. Or, perhaps you want to create a numbered outline—a task you can quickly accomplish with a numbered list. You can also use definition lists to create a "false indent" for your text—a handy hint to help you circumvent using a style sheet. You can easily program a list in HTML, in many different formats, such as: ordered, unordered, and with definitions. A few simple steps will help you meet this easy task.

Ordered List—Numbered

In an ordered list, each item is numbered and displays in a specific way. You could use an ordered list to demonstrate a series of steps in a recipe, list a set of instructions, or for many other purposes. For example,

1. Do this
2. Then do that
3. Do something else
4. Take the final step

To start an ordered (numerical) list, start by surrounding your entire list with the tags <OL> and </OL> Within the list, precede each separate item with a "List Item" tag <LI>.

HTML Tags	Result
<OL>	
<LI>Item One	1. Item One
<LI>Item Two	2. Item Two
<LI>Item Three	3. Item Three
</OL>	

You can also use the **TYPE** attribute to vary the numbering style for your lists. For example, when you use the following coding, <OL TYPE = "n">, "n" will produce the following types of numbering schemes:

Type	Numbering Scheme	Example
1	Standard numbers	1, 2, 3
a	Lowercase letters	a, b, c
A	Uppercase letters	A, B, C
i	Small Roman numerals	i, ii, ii
I	Large Roman numerals	I, II, III

Unordered List—Bulleted

An unordered list uses bullets rather than numbers to highlight the individual items on the list. For this particular list type, the order of the items is usually not that important—for example, a shopping list. To set up an unordered (bulleted) list, start by surrounding your entire list with the tags <UL> and </UL> Within the list, surround each separate item with the "List Item" tags <LI> and </LI>.

HTML Tags	Result
<UL>	
<LI>Item One</LI>	● Item One
<LI>Item Two</LI>	● Item Two
<LI>Item Three</LI>	● Item Three
</UL>	

As with the ordered list, use the attribute **TYPE** to select the bullet style for your list, e.g., **<UL TYPE = "n">**, where "n" = the type of bullet as follows:

Type	Bullet Appearance
circle	o A hollow circle
square	■ A solid square
disc	● A solid circle (default)

Definition Lists

In a definition list, each entry has two parts—a term and its definition. People originally used this type of list for dictionary-like lists, although nowadays it's often used as a format method because of its indenting capabilities. A definition list presents the defined word on one line and its definition indented underneath. There are three tags:

- ◆ **<DL>** defines the beginning of a list of definitions

- ◆ **<DT>** precedes the definition title

- ◆ **<DD>** represents the definition text

To set up a definition list, start by surrounding your entire list with the tags **<DL>** and **</DL>** Within the list, precede each separate item with the **<DT>** tag and then precede the definition for that term with **<DD>**.

HTML Tags	Result
<DL>	
<DT>Bronze Level<DD>Service provided for 3 months	Bronze Level Service provided for 3 months
<DT>Silver Level<DD>Service provided for 6 months	Silver Level Service provided for 6 months
<DT>Gold Level<DD>Service provided for 12 months	Gold Level Service provided for 12 months
</DL>	

Nesting Lists

You can mix and match your lists by placing one type of list inside another. For example, perhaps you're outlining a procedure with a set of steps. One or more of the steps may need to be further broken down into smaller steps and you can accomplish this by creating another list inside the original list.

HTML Tags	Result

<OL>
  <LI>Make Cake</LI>
  <UL>
    <LI>Mix sugar and butter</LI>
    <LI>Add beaten egg</LI>
    <LI>Add sifted flour</LI>
    <LI>Bake 20 mins at 350</LI>
  </UL>
  <LI>Make Sauce</LI>
  <LI>Serve</LI>
</OL>
	1. Make Cake
	  • Mix sugar and butter
	  • Add beaten egg
	  • Add sifted flour
	  • Bake 20 mins at 350
	2. Make Sauce
	3. Serve

Tables—Emphasize and Organize

If you're a new Web designer, you're in for a rude surprise: Tabs do not exist within HTML. Sure, you see pages that look like they're using tabs or columns, but the coding secret involves the use of graphic spacers or tables with invisible lines. Fortunately, you can work with tables pretty easily, once you know the coding structure, as defined by this simple table:

Coding Structure for 2-Column, 1 Row Table

```
<TABLE BORDER="0"
CELLSPACING="0"
CELLPADDING="9" WIDTH="780">
<TR>
<TD VALIGN="top" WIDTH="36%">
Text for column one would go here
</TD>
<TD VALIGN="top" WIDTH="64%">
Text for column two would go here
</TD>
</TR>
</TABLE>
```

Example of 2-column, 1-row table

Tables offer you the ability to establish more of a page layout for your site—from creating a nice, fully justified block of text off center on your page to creating a traditional table with fully visible lines indicating columns and rows. You can also "nest" tables within other tables. Although this may increase the load and design time of your Web page, it can create a graphically appealing result. Look at the page source for sites like www.amazon.com and www.cisco.com. Finding the table tags in those pages might prove a bit challenging, but once you identify them, you can clearly see that skillful manipulation of table tags can produce an extremely professional Web site.

Table Tags/Parameters	Description
<TABLE></TABLE>	Defines the start and end of a table
BORDER	Border size (in pixels) around table cells. 0 results in no border
CELLSPACING	Space between cells, in pixels
CELLPADDING	Increases pixel space between border and contents of cell
WIDTH	Pixel space or percentage of page occupied by table
<TR></TR>	Starts and ends table row
VALIGN	Vertical alignment; top/middle/bottom
ALIGN	Horizontal alignment, left/right/center
<TD></TD>	Defines table cell; valign and align parameters apply to this tag
ROWSPAN	Number of rows the cell spans
COLSPAN	Number of columns the cell spans
WIDTH	Pixel space or percentage of page table occupied by column

TIP: If your site shows blank space instead of your table, make sure you are not missing a **</TABLE>**, **</TD>**, or **</TR>** tag in your table code.

Frames—Why We Like Them; Why We Don't

It's probably a good time to mention frames. In effect, frames serve as subdivisions, or windows, of a Web page. Frames enable users to see several connected Web pages at the same time, seemingly as a single page. You can also design your frames to enable visitors to scroll through each frame separately, which is especially handy for long navigation lists. For example, imagine a Web page divided into three sections or frames that appear to the visitor as follows: One section at the top of the page can be used to display navigation buttons and graphics, a menu and graphic exists in the left-hand column, and the remainder of the screen can be divided vertically with some text displaying in the right-hand column.

Advantages of Framing

You may or may not have noticed frames in use as you surf the Internet, since skilled Web designers can make them appear either as very obvious sections of the Web page or as "invisible" subsections for the viewer. Since tables can create a similar effect, it's often difficult for the visitor to easily differentiate between frame or table use. Any time you notice a scroll bar that is not located on the right-hand side or very bottom of the screen, you're probably looking at a Web site with frames. Another way to determine whether a Web site has been coded with frames is to scroll up and down the right-hand side scroll bar and notice whether the entire screen changes, or just a portion of the screen. The latter result indicates frames exist on the page. Essentially, when you use frames, you are displaying multiple Web pages at the same time, therefore each frame can behave like a Web page and independently of the others; this is particularly useful if you're using one of the frames to receive dynamic content.

Frames contain certain features that make them very attractive to Web designers:

- You provide a single frame with a URL so that its contents can be loaded independently of other frames.
- If the window size changes, the frame can resize dynamically.
- You can name a frame so that other URLs can target it with content.

They can provide a good anchor point for browsing the rest of the site, since you can establish one frame as a stationary part of visitors' browser screens while they scroll through other frames and other pages of your site via the non-stationary area on their screens. This can be a useful feature if done carefully, and is often used to maintain a section of navigation buttons or a control bar, title graphics, or a banner along the top of each page. Another excellent use for frames results in a dynamic table of contents (links), usually located on one side of the page, which lets your visitors click on any topic in the table of contents and have the associated information displayed on the rest of the page. Remember, though, if you use frames, try to make them unobtrusive and without borders. Avoid making your page too "fussy" and the frames too small. Try to restrict the number of frames to no more than four or five, otherwise your Web page is going to resemble a patchwork quilt. For examples of good use of frames, check out sites like Flash Planet (<u>www.flashplanet.com</u>) and Del Rey Books (<u>www.randomhouse.com/delrey</u>).

Frame Disadvantages

Despite their benefits, frames can also cause problems. Used badly, they can make your page look cluttered and confusing. Web designers will often try to give their Web site the "look and feel" of a magazine by using many frames—the resulting frames may be so small that the visitor spends more time scrolling and navigating the site than looking at the material contained therein! Frames can also result in navigation challenges for visitors—as they click from link to link, the displayed address doesn't change—the result is confusing to some, who cannot determine which page they are visiting within the site.

Additionally, since the URL displayed in the browser navigation box points to the frameset, rather than URL of the individual frame, users who attempt to bookmark the URL or email it to a friend may encounter considerable frustration when the wrong link results. Printing becomes more difficult, as users must have selected the frame they wish to print; many Web novices may not understand this concept, even though simply clicking on the frame selects it for printing. Older versions of browsers may not "understand" frames and will not be able to display them, instead pushing out empty pages. Although this occurs more rarely now, the inclusion of the <NOFRAMES></NOFRAMES> tags will define the content to be displayed by browsers that do not support frames.

Making Frames Work

To add frames to your Web site, you will need to create several files: one to define the frameset and one for each of the pages that make up the frames. You'll also need to work with three tags: <FRAMESET>, <FRAME> and <NOFRAME>. Start with the HTML tag pairs <FRAMESET> and </FRAMESET> to define the primary file which contains the coding that identifies attributes for the other frames. You can add attributes to the <FRAMESET> tag—the most common of which, COLS and ROWS, divide the frameset vertically and horizontally into columns and rows. Inside these frame pairs, use a <FRAME> tag for each frame with which you want to work. The <FRAME> tag uses attributes to control the position, size, and other characteristics of each individual frame on the page:

```
<HTML>
      <HEAD>
            <TITLE>This is my frames document</TITLE>
      </HEAD>
            <FRAMESET ROWS= "15%, 85%">
                  <FRAME src="left1.html">
                  <FRAME src="right2.html">
                  </FRAMESET>
      </HTML>
```

The preceding example specifies the use of two frames that divide the Web page horizontally (by rows)—one that takes up 15%, and the other, 85%. To divide the screen vertically, replace the attribute "ROWS" with

"COLS" (columns), e.g., <FRAMESET COLS = "10%, 80%, 10%">. Note that if the combined values of the **COLS** attribute exceed 100%, the resulting frames will not display as you expected. To avoid this problem, you can use an asterisk "*" as a wildcard: <FRAMESET COLS = "10%, 80%, *"> This will create the three columns and fill in the remainder of the screen for the third value.

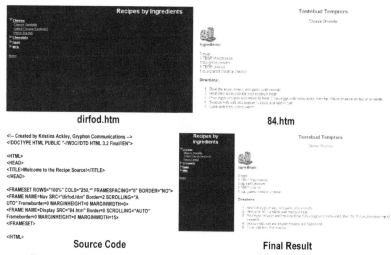

dirfod.htm   84.htm

Source Code   Final Result

Examples of frame, broken down by individual elements and coding

Next, adding content to your frames provides a slightly greater challenge than just adding text to your tags. You actually must create separate files of text for each frame (frame sources) and use a <FRAME SRC> tag to "point" to the location of those text files. In the preceding example, the two frames will display two files: left1.html sits in the first row, taking up 15% of the screen, and right2.html sits in the second row, occupying 85% of the screen. Note that the <BODY></BODY> tags should not be used for frame documents.

In case a visitor to your site is using an older browser that is unable to interpret and display frames, take some time to provide some text or content that they will be able to see. Insert content between the

<NOFRAMES> and </NOFRAMES> tag pair, such as: "Your browser does not support frames." This isn't very friendly and certainly won't encourage that person to come back to your site, so you could add a link to a no-frames version of your Web page with the following code:

```
<NOFRAMES>
        <P>Please click <A HREF="noframespage.html">
        here</A> to see the no-frames version of this
        page</P></NOFRAMES>
```

If you don't want to include a link, you can ensure that a no-frames version comes up automatically by adding some code between the <NOFRAMES> and </NOFRAMES> tag pair that displays an alternate page. For example,

```
<HTML>
 <HEAD>
 <TITLE>Using the NoFrames Tag</TITLE>
 </HEAD>
        <FRAMESET ROWS= "15%, 85%">
                <FRAME src="left1.html">
                <FRAME src="right2.html">
                <NOFRAMES>
                        <BODY>
                        ...normal Web page coding for alternative
                        Web page...
                        </BODY>
                </NOFRAMES>
        </FRAMESET>
 </HTML>
```

When working with frames, be careful when you include links within the .html documents used as frame sources—the links' behavior can be unpredictable. For example, if you're looking at a Web page that contains frames, and you click on a link within one of the frames, where does the "new" page to which you have linked appear on your screen? Sometimes, it will appear as a completely separate window (inadvertently open multiple browser windows for users). Other times, it will take over the entire window, and your carefully planned frames layout becomes a thing of the past. To solve the problem, tell the browser how to "control" the way the linked page opens by naming your frame pages: <FRAME SRC="File1.html" NAME="Left">. Now that your frame has a name, make the link target that specific frame on the page by adding the TARGET attribute to the <A HREF> tag. For example,

```
<HTML>
      <HEAD>
      <TITLE>Page of frames</TITLE>
</HEAD>
      <FRAMESET COLS="30%,40%,30%">
            <FRAME SRC="File1.html" NAME="Left">
            <FRAME SRC="File2.html" NAME="Center">
            <FRAME SRC="File3.html" NAME="Right">
      </FRAMESET>
</HTML>
```

When you add the code for the link, add the TARGET attribute and the name of the target frame. For example, <A HREF="File1.html" TARGET="Left">This is a link to File1.html</A> This example will place the contents of File1.html in the frame named "Left." If your frame contains many links, you need not go to all the extra effort of adding TARGET to all your <A> tags; simply specify a default target so that the browser can send all the linked pages to whichever frame you've named as the default. Whew! Just add the following tag to the head section of the page that contains all the links. For example, <BASE TARGET="Left"> will make all linked pages open and display in the frame called "Left." For more information and help with frames, try some of the following online resources:

Sharky's Netscape Frames Tutorial
www.sharkysoft.com/tutorials/frames
This series of short, easy-to-follow lessons takes you through the essentials of working with frames. Learn how to lay out frames, use links to populate the frames, and spawn multiple browser windows.

Framing the Web
http://webreference.com/dev/frames
Framing the Web offers an excellent introduction and tutorial on using frames. This site has a comprehensive table of contents for everything you need on frames. The "Frames Cheat Sheet" is particularly useful and well laid out.

Johnny's HTML Headquarters
http://webhelp.org/frames.html
This site has a really easy-to-follow tutorial on using frames. There are some nice examples of what happens when you use the different attributes and attribute values.

The "No Frames" Movement
www.bright.net/~jonadab/noframes
Not everyone is a fan of using frames, and this site highlights the reasons for frame avoidance. The information covers problems with using frames, as well as how to achieve almost the same results without using them. Links to additional resources discuss problems with frames, as well as links to browsers that either do not support frames or can have the frames support turned off.

Cascading Style Sheets

Cascading Style Sheets (CSS) enable Web site creators to more easily and effectively control how pages will display by defining the style instructions to be applied throughout your Web site. These are excellent tools for managing your content "cascade" because several different style sheets can be implemented together to produce the effects you seek—across different pages! In theory, they work almost like the styles you use for word processing or desktop publishing—you can actually "tell" your Web site how to produce indents, certain styles for headers, links, regular text, line spacing, margins, and much more. If you later need to make style changes to all of your Web pages at once, you can do this quickly and easily by updating the CSS. The CSS was created by the World Wide Consortium (W3C). However, at the time of this book's creation, CSS functions were not universally and consistently implemented across the newest, most popular browsers.

As this technology emerges more thoroughly on the Web, however, expect great results and more flexibility and control in your Web design efforts. In the meantime, ensure that you produce pages viewable in any browser by creating your pages first and applying the style sheets afterward. If you're really concerned about browser compatibility, you can develop browser-detecting scripts or download them from a site like www.webreference.com (search for "browser sniffer"). You can also develop different style sheets for the different browsers and their versions. And, always use linked, rather than embedded, style sheets to ensure that a single change made to a style sheet will apply to all Web pages linked to it—and the end user will only need to download a style sheet once, rather than multiple times.

> **Style Sheet**—*a document, similar to a template, which contains a collection of HTML style definitions for a Web site*

Great print resources for CSS reference include the following:

- *Core CSS Cascading Style Sheets,* by Keith Schengili-Roberts

- *Cascading Style Sheets, 2nd Edition: Designing for the Web,* by Hakon Lie and Bert Bos

- *Cascading Style Sheets, The Definitive Guide,* by Eric A. Meyer (O'Reilly Press)

To implement style sheets, you need to follow two simple steps. First, create the style sheet document (for this example, let's call it style.css and make sure you save it as a plain text file) and the "rules" that tell the browser how to display the page. Style sheets should not contain HTML tags, so avoid using <HTML> or <BODY> within your .css file! The basic formatting for a style sheet includes two parameters: the selector and the declaration. The declaration consists of two components: the property designation and the value:

Selector {property: value}

The selector identifies the HTML tag (such as **IMG, P, H1**) or another appropriate item that can be processed. The property tag tells the browser which type of style (margin, font-style, background-color, for example) to apply to the page. The value then specifies the attributes (bold, 24pt, 0.5 in) for the particular style. The whole thing is called a "style definition."

H3 {font-weight:bold 11 pt Arial, sans serif; color:pink}

The preceding rule would overwrite the existing style definitions for all level-three headers, resulting in an 11-point, bold, pink text in Arial font (or another sans serif, should Arial be unavailable). The following table provides examples of some commonly used CSS selectors, properties, and values, as well as descriptions of the end result.

> **Sans serif:** *A type of font missing the small decorative line that embellishes the standard straight-lined character.* Arial *is a sans serif font;* Times New Roman *is a serif font.*

Selector	Property	Values	End Result
H1	Color	List any color	All level-one headers appear as the identified color. This overrides all other declarations
	Background-color	Blue	Creates a blue page background
	Background-image	Fun.gif	The background image is fun.gif
	Background-repeat	No repeat	Do not repeat the background image
H1	Font-size	Can use unit (12 pt), percentage (200%), absolute (x-small), or relative (larger/smaller)	All level-one headers appear the size you have selected
H2	Letter-spacing	Choose unit of measure	Selects space between letters of level-two header
P	Font-weight	Choose any number	All paragraphs appear in the weight/boldness of the font you have selected—the higher the number, the heavier the font's weight
P	Word-spacing	Choose unit of measure	Creates white space between all words
P	Font	Font-style font-variant font-weight font-size font-family	Shorthand for all of the previous font tags combined; must be listed in order
P:first-line	Color	List color	The first line of every paragraph appears in the color you select
P:first-letter	Font-size	Choose size	The first letter of every paragraph appears in the size you select. Best used for block-like paragraphs and drop caps
P	Font-style	Italics, oblique, normal	All paragraphs appear as italic, oblique, or normal (depending on your selection)
P	Text-indent	Choose unit of measure	First line of paragraph indents
P	Margin-left Margin-bottom Margin-right	Choose unit of measure	Sets margins of paragraph according to measurement unit
Body	Background -attachment	Scroll	Scrolls background with the body of the Web page
UL	List-style-image	Bullet.gif	Will use bullet.gif as the bullet for all bullet styles

Additionally, by using class selectors, you can tell the CSS to recognize and apply specific styles to all elements identified by a specific class:

<H2 Class="special">Special text</H2>
<H3 Class="special">More special text</H3>

Then the class name would be added to the HTML selector, separated by a period:

.special {font-size : 20 pt}

The result would be the application of a 20-pt font size to all items marked as special within the document—in this case, level-two and -three headers.

Remember that second step? Now that you understand the basic CSS tagging system, it's time to insert the style sheet information into your HTML document. To do this, you'll need to place the following code between the <head> tags of your HTML pages:

<LINK REL=stylesheet TYPE="text/css" HREF="/stylesheets/style1.css">

or

<STYLE>
    @import url (http://directory/style1.css);
</STYLE>

The second example, known as "importing," will let you apply numerous style sheets to the same document, whereas the first example allows only one style sheet to link to the document. Only certain browsers (IE 4 and above is somewhat unstable with this selector; Netscape versions 6 and above support it) support @import, so check before using this tag.

One important bit of information to remember when working with style sheets: the author's rule always outweighs the users' and the browsers' rules. You can finally overcome issues like font size, background color, etc. that users set as defaults within their browsers, so long as you list those styles you want to change within your style sheets. Otherwise, if the user has no preferences set and you have not identified properties within your style sheet, everything will revert to the browser's default. To see examples of CSS in use, look through Web sites such as www.classmates.com and www.ivillage.com. For more information about style sheets, check out the following links:

Webreview's Style Sheets
www.webreview.com/style

Webreview.com pulls together an excellent list of resources referring to the successful creation of Cascading Style Sheets. Its comprehensive master list identifies everything you ever wanted to know about CSS declarations and how they are supported by the different browsers. For a quicker evaluation of how the browsers support CSS1, the Leader Board breaks down the Master List and shows the browsers' effectiveness. The site also offers a great FAQ that consistently expands on CSS information.

World Wide Web Consortium
www.w3.org

The World Wide Web Consortium (W3C) develops specifications, guidelines, software, and tools for the Web, and should be referred to for any HTML standards. Its CSS resources include a handy CSS validator that will check your codes against the latest HTML standards. It also features tutorials, a great set of links to more CSS resources, and a list of the browsers that support CSS.

6

Your Best Image

Image creation and manipulation result in some of the greatest headaches in Web design. Aspiring designers start out with good intentions, only to wind up with pictures that look like they belong on a funhouse mirror! In order to successfully manipulate images for Web site use, you'll need an understanding of the essential rationale behind image creation and modification principles, a bit of training on the program you choose, and plenty of practice (also known as "play time"). For the purposes of this book, we'll skip the lessons about the main design programs: Paint Shop Pro, Photoshop/ImageReady, Fireworks, and Illustrator. If you want to learn specifics about any of those programs, check out the sites discussed in Chapter 2, The Tools You'll Need, or visit your local (or online) bookstore to find a good book.

When working with images for the Web, you can choose several paths along which to proceed:

1. **Start from scratch.** Create your image from nothing, either by sketching it on a piece of paper and scanning it into your computer, or by "drawing" it on your screen, using one of the aforementioned programs.

2. **Use an image that exists in "hard copy" format.** This can include anything already in existence (not electronically)—a picture of a family pet, a piece of fabric, some flowers, or anything else you find lying around the house.

3. **Modify an electronic graphic.** Acquire a free graphic or purchase art from a group such as PhotoDisc (www.photodisc.com) or EyeWire (www.eyewire.com) and modify it electronically to fit your needs.

Whichever path you choose, this chapter will help you adapt the image so that it can best fit standard Web requirements. If you want more specific information about working with images, try any of the following resources:

- *Start with a Scan,* by Janet Ashford and John Odam

- *Deconstructing Web Graphics,* by Lynda Weinman

- *Website Graphics Now,* by Noel Douglas, Geert J. Strengholt, Willem Velthoven

Guidelines for Responsible Image Use

Although Web graphics may seem to provide some cool enhancements to your Web site, they can also detract from it. Negatives of using graphics include slower load time, needless distraction, and difficulties viewing the graphics on all monitors (especially for people with disabilities; see Chapter 8, Troubleshooting Your Efforts, for more information on this). To conquer these issues, set firm standards for your graphics, such as the following file size guidelines:

- Backgrounds—4k

- Logos—12k

- Image map—60k filled/30k w/text

- Custom bullets—2k

Avoid using images just for the sake of using images; this caveat especially applies when creating images that look like text. With the increasing popularity of Cascading Style Sheets (Chapter 5, Code Challenges), you no longer need to restrict your text choices to a couple of fonts. Save some bandwidth and don't create images that serve only to show a different font style!

Image maps—*a single graphic image that contains many "hotspots," or sections of the graphics which hyperlink to other areas*

Additionally, many people turn to image maps as navigation tools for their home page. Make your image maps specific and intuitive—clearly separate and label different portions of the image so users can identify which areas link to other sections. Additionally, consider a server-side versus client-side image map. Since the server hosts the image map, rather than the user needing to load it, the user can see a URL for each individual element in the image map.

You can change the size of the graphic using your Web program; however, unless you're using a design tool like GoLive or Dreamweaver that links to a graphics program, you're liable to greatly reduce the quality of your graphic, sometimes resulting in a very pixelated or blurry image. Additionally, changing the code, rather than modifying the graphic properly, can result in a graphic of poor quality. Instead, take an extra few minutes to produce a high quality image through a reputable graphics program like those described in Chapter 2, The Tools You'll Need.

Measuring Up: A Briefing on Graphic Terminology

The first step to understanding Web graphics involves a basic grasp of the terminology you will encounter. The core terms center on measurement, such as dots/pixels. Your screen consists of lots of dots, and when you adjust your monitor settings to a higher screen resolution, you set the number of dots that will display on your screen. For example, a screen resolution of 800 by 600 pixels actually produces 800 dots per each of 600 lines, resulting in 480,000 pixels.

As you move into larger monitors, the number of dots per inch (dpi) decreases, meaning that the clarity also decreases. For example, a 15-inch VGA monitor might display 51 dpi, whereas a 19-inch monitor could display only 22 dpi. Of course, the higher a dpi number, the better the graphic quality, as the concentration of dots increases and higher degrees of graphic clarity can be obtained. Unfortunately, since monitors only go up to a certain dpi level, you really won't create better Web graphics when you select higher resolution sizes for them. Most printable images should use 300 dpi; Web images should be set for a 72-dpi resolution.

original picture

Close-up of adjusted image shows jagged, "pixelated" result, rather than the original's smoothness, resulting from poor adjustment.

Example of poor adjustment, creating a pixelated effect

However, when you work with images outside of the Web, you'll want to save them at a higher resolution to ensure better quality as you manipulate them. If you attempt to enlarge a GIF or JPEG sized to 72 dpi, you will create a pixelated image that looks very amateurish, with jagged edges. Therefore, start with a high quality image (such as a .tif or .psd) and spend your time perfecting the image in a true graphics program. You can choose to work with rasterized images, which are recorded by pixels to display, and can only be edited by changing the pixels in a bitmap editor like Photoshop. Or, you can work with vector images, which have been recorded by geographic shapes, though they display on your monitor as bitmap (GIF and PNG, described shortly, are vector images). Although you can more effectively resize and rotate vector images, you will have a tougher time finding a program to successfully handle this task for you. After you have modified the image to your satisfaction, you will compress it via the tools available for that particular file format, and save it as a 72-dpi Web image.

Types of Images and Best Uses

When it comes to Web design, you may initially work with a variety of graphics (such as .eps, .tif, .psd, and .ai). Use these high quality images as the baseline from which to develop your Web graphics, since they offer higher quality and more flexibility than standard Web graphic formats. In the end, however, you need to create two standard types of graphic files: GIF (pronounced with either a hard "g" sound or a soft "j") and JPEG (jaypeg). A third type of graphic, PNG (Portable Network Graphics), looked as if it would gain popularity several years ago; however, it fell flat and only certain browsers can support it. For more information about up-and-coming graphics formats, like PNG, keep an eye on the Internet and relevant print matter.

GIF (Graphic Interchange Format)

The first file format supported by Web browsers, GIF (Graphic Interchange Formats), still remains one of the most popular graphic formats for the Web. Both to their advantage and disadvantage, GIFs work from an 8-bit palette, which enables you to save them as small, highly compressed image files, but they can contain only up to 256 colors. GIFs also offer several additional, interesting attributes: interlacing and transparency, covered later in this section, and the ability to animate images, as discussed later in the chapter. They are excellent file types for flat images (such as cartoons, drawings, and other forms of "line art"), since they compress color information according to pixel rows, meaning that they treat similarly colored pixels as single units.

GIFs compress files very efficiently, through lossless, LZW (Lempel-Zev-Welch) compression.

Line art—*a graphic that consists solely of lines, without any shading. Photographs and many computer images include shading so cannot be considered line art*

Lossless means that none of the original image information gets "lost" during the image compression, resulting in a smaller image identical to the original image, unless you change the colors. LZW compression helps GIFs identify the repetition of color, efficiently minimizing the number of pixels stored by color. For example, a row of 10 pink pixels would be stored as "10 pink." You lose the compression advantage, however, when you try to save a row of 10 pixels that fade from red to pink. Since the colors do not repeat, the resulting GIF would require more storage area, or you need to minimize the color change in that row, which could be difficult when working with photographic image.

Interlacing GIFs: Blessing or Curse?

Traditionally, GIFs appear on your screen a bit slowly, displaying alternating rows one by one, from top to bottom. You can make image loading more interesting for your users by selecting "interlacing" as a different method. Interlacing cause an entire image to quickly display on screen, then the details slowly fill in, providing visitors with a sneak peek of the graphic to be loaded. This can prove especially helpful when using larger images, enabling the user to switch to the next page, rather than waiting for a graphic they do not want to see (such as an image map). Conversely, interlacing can increase your file size, and some graphic purists would rather see nothing at all than a partial, blurry image.

Transparently Thinking

GIF's other unique feature, transparency, addresses some previous issues related to the quality of the graphics. In the past, images that did not look like a traditional, square shape would end up needing to fill the remaining space in the rectangle, via some color, causing a less integrated look on the page. By incorporating a transparency effect, via the GIF89a format, you designate the extra color surrounding your primary image as "transparent," meaning that that color will not show up against its background.

Transparency can prove a difficult task. First, you can only select one color to play the role of the transparent color. Second, ANY and ALL pixels in that particular color will turn transparent. Therefore, if you work with a cartoon showing a bouquet of red and yellow flowers, for example, and you want the space between the stems to appear transparent, rather than white, against your background, you need to follow several steps. First, choose a color (gray, by default) which you want to make transparent. If you select red, all of your red flowers would turn clear, so make sure you do not use a color that appears somewhere else in your image. Next, apply the transparent gray color to any section of your image you wish to make transparent.

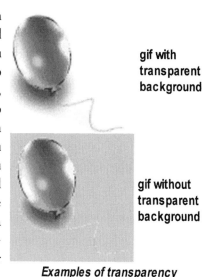

gif with transparent background

gif without transparent background

Examples of transparency

Minimizing the Size of Your GIFs

Despite the fact that GIFs can produce very small file sizes, sometimes they need a bit of assistance from you to achieve that goal, and exceed it more than you thought possible. Fortunately, most of the image editing programs contain some specialized methods for decreasing the file size. However, you may need to exceed their standard offerings and experiment with options like dithering and color depth reduction. Dithering results when a dot pattern varies to create an illusion of new colors and shades. It will make your GIF images look smoother and decrease your chance of jagged edges, but will also increase your file size. By reducing the amount of dithering, you can decrease your file size, but you also take a chance with your image quality, so carefully examine your attempts.

Next, take a careful look at the colors used in your graphics. Can you remove some of them without affecting the quality of the graphic? You can quickly determine this by reducing the number of color bits; start by reducing the number of colors from 256 to 128. If the image still looks great, try a lower color depth, and keep trying new ones until you reach the lowest possible. If you go too far, take a step back and undo your efforts. This technique can also apply to JPEG files.

JPEG (Joint Photographics Expert Group)

JPEGs, on the other hand, work extremely well with photographs and other images that use complex shading, as its file format will compress these graphics more thoroughly. You can also work with a greater range of colors; JPEGs support 24-bit color palettes, which result in millions of colors. Rather than recording the color for each pixel (or row of pixels), JPEGs will average the different hues, creating more subtle transitions between colors, which our eyes interpret as smoother, higher quality images.

Additionally, most graphic programs compress JPEGs well enough when you output them that you will not need to concern yourself with minimizing them. You can apply some of the optimization efforts described in the GIF section, but pay careful attention that you do not cause artifacts to occur. Artifacts result in portions of the image clumping together, creating a lumpy, bumpy look to your graphics. Also, pay close attention to the way your JPEG file works with sharp edges. GIFs work much more efficiently with lines and corners that do not fade; JPEGs may produce "jaggies," or a staircase-like effect.

Unfortunately, JPEGs contain some negative results, despite their obvious positives. Most obviously, you cannot create a transparent JPEG. First, they use "lossy" compression, which results in some sections of the images being discarded (or lost). You may not notice this effect in high quality, photographic images; however, it will produce a very mottled looking image. Also, on a system set to produce less than 24-bit color palettes, the colors in the JPEG will automatically reduce, resulting in some shading (dithering) effects. Though you will not notice this result as strongly with JPEGs as GIFs, you should still pay careful to the quality of your image.

Additionally, JPEGs can cause some serious issues when you try to edit them. Although they can support resolutions over 72 dpi and work well for print production, as well as electronic, you should avoid resizing and resaving the image, because you cannot regain the file quality once you have changed it. Therefore, you should always save a copy of your original JPEG file, or work from a higher quality image type, like .tif, before saving your final efforts for the Web. Despite the negatives highlighted in this section, JPEGs should be used to support the high quality demanded by photographs and other similar images.

Start with a Scan...or a Digital Photo

As mentioned earlier, your image creation process can start with non-electronic images. Scanners and digital cameras offer two quick and easy ways to turn real, physical data into an electronic format. A few years ago, these products were so expensive that few people other than graphic production artists and large companies had the financial resources to afford them. Fortunately, while the quality of and demand for these tools continue to increase, the cost has decreased. More and more people are purchasing a high resolution digital camera or using a combination of their traditional camera and a new scanner.

The standards for each constantly change, so be sure to check out resources, such as Computer Shopper (www.computershopper.com) for more insight on the current standards. You can also consider purchasing a previously owned scanner or camera. Many people who seek to own the most technologically advanced items will return fairly up-to-date equipment that you can buy for an extremely low price. Additionally, major companies like Hewlett-Packard (www.hp.com) sell refurbished machines that contain the same warranty and quality guarantee as original equipment.

Best Practices

Make sure you learn how to adjust your scanner or digital camera to produce the highest quality original image, and aim for 16 million colors whenever possible. Although this may take some extra disk space, you will experience greater flexibility when you attempt to modify the image. You will always experience better results if you decrease the dimensions of an image, rather than increasing it. Increasing the size can cause the quality to deteriorate, resulting in a pixelated picture. However, if you know the final dimensions of the image you wish to produce, scan in the image at that size—you will always produce the best results when you know the exact dimensions.

To increase the quality of a scanned image, make sure you position your item in a perfectly straight manner. Although this may take several attempts, you will produce better quality with an item initially set straight than working with your graphics program to rotate the item. Additionally, always work with an original document. Have you ever seen a photocopy filled with crumpled lines that has been partially rotated on a page? Most likely, the person responsible for the photocopying did not work with an original document. Don't cause the same effect with your scanned documents. Although you can spend time cleaning up the image later, start with high quality, camera-ready material for the best result.

Mickey Mousing Around—How to Work with Animated Images

Now that you know how to produce high quality, static images, consider venturing into animated images, an area known for its fun, dynamically creative aspects. Animated images that spin, whirl, and dance across the screen can really bring a touch of magic to an otherwise basic Web page. Incorporating animated images into your Web site will catch the eye of the surfer, offering an excellent way of adding definition to specific elements that you may want to emphasize, such as a graphic that offers a special discount.

Take care not to overzealously add animations to your Web site, however. Though they lend liveliness and pizzazz to your pages, they can also overwhelm your users, not only through their constantly changing aspect, but also through their file size and, therefore, load time. More than just one or two animations on a page can distract and annoy your site visitors, as they attempt to determine the most important data. Animations can also serve as a source of great irritation to visitors with lower Internet access speeds, especially when they appear on your front page. Animation at the beginning of a Web site that takes forever to load may turn away some users, since in the world of cyberspace a few seconds can result in just a few seconds too many for surfers in a hurry.

You can create your own graphics using one of the many drawing programs available, such as Photoshop or Paintshop Pro. Follow the steps listed in the GIF section of this chapter to create and perfect your .gif image or series of images. The next steps for animation will actually create variations—each one slightly different—of your original file and incorporating them into a single, animated image. By using a program that can animate your images, you specify that the images be displayed sequentially, thus creating animation—rather like the old-fashioned "flip books" that people used to quickly flick through to see the pictures morph into a moving image. Make sure that all of your images are the same size and that you change the color, shape, or text slightly each time, in order to create the effect. Once you've created your animated image, put it on your Web page by using the <IMG> tag with the SRC attribute pointing to the animated file.

Macromedia's Fireworks program and Adobe's Photoshop and LiveMotion programs can provide excellent resources for your animation needs. If you want to find less pricey shareware or freeware programs, take a look at the animation software available on the Web. Sites particularly helpful in your software search include:

Alchemy Mindworks Inc.
www.mindworkshop.com

The Alchemy GIF Construction Set, a popular GIF animator, is a powerful tool to help you create animated GIF images. This site also offers some great animation resources, including books, articles, and relevant links.

Lycos Computers
www.lycos.com
Use the search capability within this site to search for GIF utilities. Select GIF animators and take a look at the shareware section that allows you to view a list of reviewed GIF animators. Once you decide which one to try, you can download it or go to the software's home site to get more information.

Animation World Network
www.awn.com
A portal of animation resources, the Animation World Network is updated weekly and includes an "Ask the Expert" forum and software reviews.

On the other hand, if you're experiencing a lazy day and don't want to go to the trouble of creating your own animated images, you can download them from one of the many Web sites that offer free or reasonably priced images, as described in the next section.

Freebies—Where to Find Free Images Online

Luckily, if you do not want to create your own images, you can access literally tens of thousands of free graphics available to you at the click of a mouse. The bad news is that you need endless patience and finely honed sleuthing skills to really hunt down the great sites. It's literally a jungle out there in cyberspace, but with a little bit of patience and persistence, the Web will lead you to a treasure trove of clipart wonders. During your search, you may find many of the free clipart and graphics sites laden with those annoying pop-up ads, animated ads, links that lead to more ads, etc. You can experience a nightmare trying to navigate through a veritable minefield of annoying interruptions. If you're not careful, you'll find yourself clicking on what looks to be a promising link, only to find that instead of being able to view some nifty graphics, you're hyper-jumping to another Web site, which in turn takes you to another, and another, and so on. You can eventually give up all hope of actually reaching some bona-fide clipart!

On a good clipart site, you'll usually see a table of contents showing the available graphics. In addition, you may find a list of recommended or rated Web sites that also contain free graphics. Some Web sites are nothing more than portals to other sites. The worst types of clipart sites exist only to redirect you to another Web site. If you're unfortunate enough to wander into one of these, you may start to worry that all clipart URLs lead you on a wild goose chase; rest assured that this is not the case.

Once you've found some great free clipart, which you will, make sure you read any disclaimers or copyright information that may be posted. There may be restrictions on how you can legally use the clipart image that you download. For example, many copyright notices specify that you may only use the image for your own personal use, and not for any public display. Make sure you carefully note and follow the legal guidelines, or you may encounter serious problems later (for more information about legal issues about which you should take note, refer to Chapter 1, Getting Started).

To download the clipart, simply right-click your mouse and save the image to your computer. If you're using a Mac, click and hold the image you want, move down the menu to "save image as," and release the mouse button. Most sites will also provide basic instructions on other methods you can use to download individual images on the site or a collection of images which you find interesting. We've listed one or two Web sites below to get you started. Moreover, don't forget—you can find animated graphs, as well as backgrounds, buttons, drawings, and photos, among the treasures of the Internet.

All Free Original Clipart
www.free-graphics.com

All Free Original Clipart highlights a good selection of clipart, animations, buttons, banners, fonts, backgrounds, and more. The left column of the page lists clipart categories from which you can choose your images. In addition, there is some excellent information on bandwidth theft. The

"What's New" section keeps you up to date on the latest additions to the site—a nice touch if you visit regularly.

AAA Clipart.com
www.aaaclipart.com
This Web site provides a wide selection of images and lots of links to other great Web sites that have free graphics. There's an easy-to-read alphabetical list of clipart categories. Make sure you read the disclaimer and terms of use.

Animation Factory
www.animfactory.com
Animation Factory offers an excellent source of many animated images that you can download free for your personal use. The site also features an Animation of the Day, as well as a design studio that contains an excellent section of tools and graphics components for you to download. For access to over 100,000 royalty-free animated graphics, you can join the Premium Gold program or purchase a CD-ROM that allows you to use the animations for other than personal use.

Bandwidth Bandits

Many Web sites are becoming increasingly concerned these days with Web designers known as "Bandwidth Bandits." Any time that you use HTML code to link directly to a graphic on someone else's Web site, e.g., <IMG SRC = http://domain.com/picture.gif>, instead of copying that graphic to your own Web site, you are guilty of being a "Bandwidth Bandit." This produces a drain in the resources of the site from which the graphic originates, since every user who opens another site with the graphic on it essentially causes the graphic to load from the original site. Secure permission to reproduce the graphic and place it in your site's directory, to assure that you do not transgress in this manner.

7

Special Effects

I f the Internet and all of its possibilities excite and interest you (which they probably do, or you would not have purchased this book), the multimedia capabilities of sound, video, and animation will thrill you even further. Gone are the Web's early days, when budding Web designers took their first, hesitant steps into cyberspace with Web pages that displayed text and photos and very little else. Now, there are all kinds of fascinating gizmos that you can add to your Web page to make it a true communications medium of the twenty-first century. It's not difficult to enhance your Web site with sound and movement, and a few special effects can go a long way towards making your Web site really stand out on the Internet. Make sure, however, that you use them sparingly—you don't want your visitors to experience sensory overload!

Many Web designers like to include various special features and effects described in this chapter on their Web sites because visitors to their site will enjoy a greater level of interactivity. Compare a book with television. When you read a book, you cannot interact much more than turning the pages in a specific order—one after the other, from the first page to the last—and to read the printed words and look at pictures. With television, the pictures move, you hear sound, and you can change the channel to view information completely unrelated to the previous channel. With a well-designed Web site, you'll not only find text, but also perhaps color, graphics, moving images, sounds, and interactive tools that let visitors jump around from place to place, play games, or otherwise take action on the site. To many people, the different "bells and whistles" on a site create a dynamic, fun atmosphere to which they will want to return numerous times.

10 Things That Can Make Your Web Site More Fun

1. **Animation**—adds life to the page, without overwhelming

2. **Video**—keep it short, simple, and relevant

3. **Audio**—add another dimension to your page

4. **Clickable elements**—hop around the site

5. **Instant/automatic response**—everyone hates to wait; don't make them

6. **Interactive forms and other stuff**—get your users involved

7. **Search engines**—find what you seek

8. **Graphics**—take a step beyond plain text

9. **Color**—enhances and generates feeling

10. **Hit counters and guest books**—find out what your visitors really think

Some of the above are covered in other chapters; the rest we'll cover in this chapter. Many of the multimedia effects, such as video and sound, will require your visitors to download special software called plug-ins or players in order to see and hear them effectively. This can prove daunting to some Web site visitors, even though it's really not complicated and merely requires the download and installation of software. See the basic steps for downloading plug-ins at the end of this chapter; you should always make plug-ins easy for users to download if elements on your site require them. Other special effects include little extras such as hit counters, search boxes, Java applets, and Flash animation. Used thoughtfully, they can all help to elevate your Web site to the dizzy heights of advanced design.

Plug-Ins—*software modules that enhance and extend the capabilities of your browser, enabling it to support specific applications or file formats*

Hit Counters and Guest Books—Interacting with Your Audience

Now that you've built your Web page, do you care if anyone ever visits? Of course you do. You're proud of your work and you have something to say, so naturally you're curious to know how many people have checked it out. Therefore, you might consider adding a hit counter (also known as an access counter or Web counter).

Hit counters collect statistics concerning the number of times a Web page is accessed, and can generate that information through periodic reports. Usually, the hit counter is placed at the bottom of the Web page and displays the numerical results through whichever graphical format the designer has applied. You've probably seen plenty of them on Web sites—quite often, they're small rectangular boxes accompanied by the text "You are Visitor number xxx." Other times, they may appear as elaborate design elements, with animals or flowers winding around the numbers. The possibilities are endless; you just need to choose the format from the variety of excellent Web sites offering them, and apply them to your page.

Sometimes, you may encounter a small challenge when you attempt to apply a counter to your site. Hit counters come in many different languages—from JavaScript to CGI and beyond. Some of the free counters may require a small advertisement be applied near the counter, but they will track the results and you need only cut and paste code into your HTML document. On the other hand, others may need more technical expertise, and will require you to work with your Web site host to track down the information you need.

Additionally, you should know some of the downsides to counter use. They can definitely give your site visitors a negative impression—if your displayed counts are low, some people may leave as a result, thinking your site is not "worthy" because of the low number of hits. Some Web site designers program the counter to start at a specific non-zero number, which means that they have given themselves a "head start" in seeming to be well-visited and popular. Additionally, some designers consider visible hit counters to be a little amateurish; you can easily remedy this by using a counter that does not show up on the Web page—i.e., one that is transparent or "invisible." Finally, hit counter statistics do not provide the most accurate measurement of site usage. They normally track the number of the page accesses, not the number of pages accessed by different visitors. If you want an accurate determination of your site's success, through the measurement of traffic, you should consider some of the tracking tools available on the market, such as WebTrends (www.webtrends.com) or SuperStats (www.superstats.com).

Of course, like everything else, you can create a hit counter program yourself, or you can simply download one from the Internet. Often, the sites will give you the HTML code to insert into your Web page source code, but you will still need to coordinate the script hosting through your Web host. While some counters may require a nominal fee, many of the basic ones are free. Be aware that some free hit counters are combined with ads from the sponsors—after all, they benefit from advertising on your Web site, and you benefit by getting a free hit counter! Some sites to check out for hit counter information include:

Cranfield University
www.cranfield.ac.uk/docs/stats

If you've ever wondered what you can do with your hit counter statistics, this site has information on Web statistics—the good, the bad, and the ugly. The section on "Quick Questions and Answers" really covers all the essential questions that you may want to ask. There's also a feedback form if you need to add a question.

Microsoft Fast Counter
http://more.bcentral.com/fastcounter

This site offers you a free hit counter. Choose from several styles, and installation is quick and simple—just five minutes. With this free service, you'll receive emails of stats updates, in exchange for placing a link to the Microsoft bCentral site somewhere on your Web site.

The Counter.Com
www.thecounter.com

Choose from a selection of free hit counters and reports. This site helps you install the counter, as well as providing some great resources, such as FAQs, articles, featured sites, and "The Web Developer Channel." The helpful, daily, in-depth traffic reports will track the number of visitors—just add a few lines of HTML to your code. You can choose from several different styles of hit counters, including an invisible counter.

Live Counter Classic
www.chami.com/counter/classic

You can download a free hit counter from this site. Its more popular choices include an animated odometer-like display that will increment as you watch. Alternatively, you can choose to hide it on your Web site. Your hit counter can also count visitors from non Java-enabled browsers.

Knowing how many times your Web page has been accessed provides just one piece of useful information. But, how do you know what your visitors really think about your site? Knowing something about your audience can be extremely useful in continually improving your Web site to attract more and more visitors. Accomplish this by adding a guest book to your Web site. A guest book not only helps you to get visitor feedback, but also adds an interactive element that makes your Web site more exciting. Visitors to your site can post their own comments, as well as read comments from other visitors. However, keep in mind that anyone can post anything to a guest book—you should keep an eye on the comments made, as some irresponsible visitors might post inflammatory or inappropriate material.

You can design your guest book to blend in with the rest of your site. Some Web sites that help you to create or download free guest books include:

Pathfinder
http://guestbooks.pathfinder.gr
Create a free guest book for your Web site. In just three easy steps, this site will help you set up a guest book for any URL. The "Guestbooks Assistant" will help you with any questions or issues that need to be addressed. There's also a nice "Guestbook of the Week" feature that showcases a selected guest book.

Bravenet Web Services
www.bravenet.com
This site provides an excellent resource for downloading free guest books and other free tools, such as hit counters, chat rooms, email forms, and a cartoon of the day. You can register at no cost and join over 1.5 million registered users of this service. Customize your guest book and incorporate many features, such as implementing checks for spelling and profanity, prevention of image posting, email notification, colors, graphics, music, and more.

Yahoo's Guest Book Index
http://dir.yahoo.com/Computers_and_Internet/Internet/World _Wide_Web/Programming/Guestbooks/
This is Yahoo's list of sites that offer guest book services. You can choose from plenty of samples and ideas for incorporating guest books into your site. Some suggestions and guest books are good, while others are lower quality—but many are free.

Java: More Than Just Hot Coffee— How Java Applets Can Enhance Your Web Site

Give your Web site a jolt of caffeine and make it dynamic and interactive with a sprinkling of Java applets. All you need is a Web browser that knows how to interpret them, and you're ready to go. You're probably quite proud of your Web site by now. If you've followed the suggestions in this book, your site should look good and you'd like to solicit feedback from your visitors. Perhaps you're creating a corporate Web site that needs a form for applicants to apply for job openings, or maybe you developed an e-commerce site on which you sell Beanie Babies, for which you need ordering and payment information from your customers.

Now's a good time to briefly review the differences between **Java**, **JavaScript**, and **Java applets**, since you'll no doubt come across references to all three. **Java** is an object-oriented programming (OOP) language that is often used to add advanced capabilities to Web pages. Java can be run on virtually any platform and is similar to C or C+, with built-in safety features that prevent some of the problems often encountered by client machines, such as the introduction of computer viruses. With Java came the capability to add sound and animation to your Web page, while allowing the visitor to interact with your site—for example, by playing interactive games. Essentially, when you access a Java-enhanced Web site, your browser receives not only the Web page but also a Java program. If your browser is able to recognize the Java program (and most do—Netscape has been Java-aware since 2.0 and Explorer since 3.0), it runs the program directly on that Web page, allowing you to play a game, use a form, or otherwise interact directly with the site.

The majority of Java that you're likely to encounter will be in the form of Java applets, described later.

JavaScript should not be confused with Java. Whereas Java, being a "proper" programming language, requires the use of a piece of software called a "compiler," JavaScript consists of different scripts, which are interpreted and run by the browser. Unlike Java, which must be created

Object-Oriented Programming (OOP)— *a class of programming languages and techniques based on the concept of an "object"—a data structure packaged with a set of routines, called "methods," which operate on the data*

and edited through a special program, JavaScript can be written using any text editor or any other editing program that you prefer. JavaScript extends the capabilities of HTML and is often used for less complex tasks such as validating HTML forms. You'll find that the majority of special effects, such as pop-up windows, alert messages, and rollover effects, are written in JavaScript. You can recognize JavaScript in Web page code from the tags <SCRIPT language="JavaScript">. There are many sites on the Internet that offer free JavaScripts that you can use on your Web site. Most of them will come with instructions on how to add them to a Web page; check carefully to make sure that you can use the script for free, as some may require payment. Sites that offer JavaScript goodies include:

The JavaScript Source
http://javascript.internet.com
This great site for JavaScript includes resources such as free scripts, reviews, and news. The "Latest Additions to the JavaScript Source" lists new JavaScripts under categories like games, navigation, forms, and calculators. "Our 10 Newest JavaScripts" is just one selection item on a pull-down box packed with goodies.

JavaScript.Com
www.javascript.com
This site describes itself as "the definitive JavaScript resource" and definitely has an extensive repertoire of JavaScripts, links, newsletters, articles, and tips. A drop-down menu enables you to select from a long list of "tip of the day" gems, and there's a useful tutorial on adding some Object-Oriented Programming techniques to your JavaScript.

Java applets refer to Java programs that are generally fairly small in size. Think of them as mini applications, or programs that can be downloaded and included in a Web page, just as an image can be included. When you use a Java technology-enabled browser to view a page that contains an applet, the applet's code is transferred to your system and executed by the browser's Java Virtual Machine (JVM). You don't have to think about installation or setup, because your browser takes care of that for you. Once you've scoured the Internet for a Java applet that you like, you can usually download it to your computer. Then add it to your Web page, just as you would call up an image or sound file, by using special HTML tags and attributes (unless, of course, you're using a Web development package that does it for you).

When you add a Java applet to your Web page, you use the <APPLET> tag. For example,

<APPLET CODE="file" WIDTH=X HEIGHT=Y>some text for a non-Java browser so that your visitors with non-Java browsers can at least read something</APPLET> e.g.,

<APPLET CODE="chessgame.class" width=100 height=130>If you could see this, you would see a chessgame</APPLET>

The "file" refers to the name of the "class" file that contains the code. In object-oriented programming (OOP), you create definitions for the types of objects you use in your program. A "class" can be thought of, very loosely, as a noun—something that identifies a category; e.g., the category "house." As a class, "house" can be assigned attributes, or features, such as floors, walls, ceilings, windows, and doors. Later on, in your program, you can define a bungalow, or a mansion, both belonging to the general house class, but in each case you can get more specific about how many floors, doors, etc., each type of house has. This is all fairly basic and general, and if you're interested in learning more about OOP, check out:

Compiler—*a computer program that translates another computer program written in a high-level programming language, into machine language so that it can be executed. The computer program being translated by the compiler is called the source program; the generated machine language program is called the object program.*

Don't Fear the OOP

http://sepwww.stanford.edu/sep/josman/oop/oop1.htm

This fun site caters to three levels of visitors: basic interest, really interested in OOP but not ready to code Java, and daredevils. This light-hearted, but nevertheless informative, tutorial presents a kind of "paint by numbers" method when explaining the OOP intricacies.

The Journal of Object-Oriented Programming

www.joopmag.com

Check out this site for how-to articles, reusable sources code, and programming techniques. Probably best for more experienced programmers, this site has many in-depth articles on a variety of OOPs-related topics, as well as a mentoring and training list of OOP resources.

In the earlier example, the applet file that's being included in your Web page is the "chessgame.class" file. This represents the really important part of this tag, since all Java applets use a .class extension as their filename. The width and height attributes specify the dimensions of the space the applet will take up on your page (in pixels). For example, the chessgame will be displayed somewhere on your page, so you need to specify how large it should be. If you download the applet from the Internet, it will usually include instructions that indicate the correct size and width. If you do not specify the height and width of the applet, it may not appear correctly on your Web page. The additional text between the starting and ending tags helps visitors whose browsers cannot read and execute Java applets identify the information they are missing. If the page is viewed by such a browser, it will ignore the APPLET tags, and will display only the text between the starting and ending tags. Most Java applets will run "as is," but you can customize others, specifying attributes like colors or content. For these, you can add additional parameter tags, <PARAM> and </PARAM>, along with your custom values. For example, <PARAM NAME="name" VALUE="value">, where "name" is the name of the parameter and "value" is the customer value that you want to use, e.g.,

```
<PARAM NAME="bgcolor" VALUE="green.blue.white">
```

Types of Java Applets

There are almost as many types of Java applets as there are flavors of coffee. You'll find them in a variety of sizes, from small animations known as "dancing baloney," to more complex, interactive forms or games. Some types of Java applets that you'll easily find to download onto your Web site include:

- Interactive games

- Chat rooms

- Message boards

- Discussion forums

- Clocks

- Hit counters

- Polling forms

Resources for Free Java Applets

Java applets provide an especially effective solution for low-tech designers, because you don't have to create them yourself, unless you want to. Just like clipart, you can use the Internet to locate many free, useful Java applets, from stock quotes to world clocks to polling forms, and more. Games and interactive forms represent the most popular types of Java applets. Just download them and drop them into your code to create a truly interactive Web site AND impress your users with the technical level of your programming.

The Source for Java Technology
http://Java.sun.com

As the official Java source, Sun Microsystems provides plenty to keep you busy. If you already know Java, you can contribute to the Java Developer Connection and Community Section via forums and the "Java Live" section, real-time chat sessions on specific topics at designated times. If you know nothing about Java, take their Language Essentials Short Course—it's online, it's well designed and informative, and, best of all, it's free.

Freewarejava.Com
http://freewarejava.com

This excellent site provides links to all things Java—books, Java applets and JavaScript, forums, and even Java-related jobs! The "What's New" page is a nice feature that lets you keep track of the latest and greatest additions. Additionally, the variety of tutorials cover topics from basic to advanced JavaScript, as well as Web building.

Java Boutique
www.javaboutique.com

At Java Boutique, you can select from an extensive list of Java resources, including applets, tutorials, reviews, and articles. The site's "Top 15" is a list of goodies that you can first preview and then download and use on your Web site. If that's not enough for you, the "Top 100" should keep you busy.

JavaScript Tutorial for the Total Non-programmer
http://webteacher.com/javascript/

This great site helps you pick up the basics of JavaScript. In addition, as the title implies, you don't have to be a programmer to understand it.

Sounds—Filtering Out the Noise

The Web hustles with the sound of music…and speech…and poetry…and special effects. In fact, the Web holds the potential of being a very noisy place if you're not careful. Multimedia is extremely popular, and by now you've no doubt experienced at least one Web site that comes up in a blaze of sound that totally startles you and anyone else close by. Sound effects on the Web can be a wonderful thing and can add value to your content, but sound for the sake of sound results in senseless noise, and most surfers won't thank you for it.

Let's say that you've decided to add a sound file to your Web site—perhaps a short sound bite to accompany an animation or video clip, a musical intro when your Web site loads, some verbal instructions, or sound effects to emphasize the theme of your site. There are several different audio file formats that you can use, and, of course, it's always best to use the most widely supported ones. Some browsers contain native support for a particular type of file format and can handle playing them without additional software. Other file formats require that visitors to your Web site use specific plug-ins in order to listen to (or in the case of video, view) the files. By the way, if you are providing a multimedia file on your Web site, it's considered good form to offer your visitor a link to a site from which they can download the required plug-in.

Popular sound plug-in software includes Windows Media Player and RealAudio. For multimedia files that contain both sound and video, RealPlayer, Flash, Shockwave, and QuickTime are popular. TrueSpeech is a player for Web browsers that plays back voice recordings. Free players are available for these formats. You can download a free version of the TrueSpeech player from this site: www.dspg.com/player/main.htm Other popular plug-ins can be downloaded from the following sites:

Player—*a type of plug-in that extends the capabilities of your browser to support specific applications or file formats*

Real.com
www.realaudio.com

You can download either the basic version for free and be occasionally plagued by advertisements for the full version, or pay a modest fee for the full version. RealPlayer gives you great audio and full-screen video.

Shockwave.com
www.Shockwave.com

Download the Macromedia Shockwave player from this site in just three easy steps. The Macromedia Shockwave Player 8 includes the Macromedia Flash Player 5.

Frequently Used Sound File Formats

- μ.-LAW (.au/AUdio): As the standard Unix format for audio, .au support is included in both Internet Explorer and Netscape Navigator. It receives wide support among browsers and serves as the standard audio file format for Java. However, as the amount of Unix access to the Web has decreased, so has the use of .au files.

- WAVE (.wav/WAVeform): Developed as the standard audio format for Windows operating systems, .wav has expanded in popularity and has gained support by being included on Mac systems, as well. The sound quality is equivalent to .aiff files, and frequently appears as 8- or 16-bit audio data. It's supported by all versions of Internet Explorer and by Netscape Navigator version 3.0 and up. This format is generally used for short sound clips.

- AIFF (.aiff, .aif, .ief/Audio Interchange File Format): Originally created as the standard audio form for Macintosh, .aiff files have expanded their horizons and are now supported by Windows, as well as other platforms. This common sound file encodes audio data in 8-bit mono or stereo waveforms.

- MIDI (.mid/Musical Instrument Digital Interface): MIDI (pronounced "middy") represents electronic music that's been created using a MIDI synthesizer. This expandable format includes values for the note's pitch, length, and volume, and can also add additional characteristics, such as attack and delay time. It's typically used for instrumental music files and is supported by Internet Explorer and by Netscape Navigator version 4.0 and up.

- RealAudio (.ra): Real Audio, the standard for real-time audio on the Web, is great for playing long, fairly uninterrupted pieces and is frequently used for live broadcasts. This format is used with the RealAudio Player plug-in.

Working with Multimedia on Your Web Site

In order to create and add multimedia (sound or video) files to your Web site, you must successfully complete four major steps: capture, edit, process, and publish. **Capture**, the first step, requires that you acquire a sound that you like. For example, you may decide to record your child singing, your dog barking, a short piece of music that you like, or a particular sound effect. Recording can be accomplished as a one-stage process, by recording directly from the actual sound to the computer (direct-to-digital recording), or as a two-stage process, by recording first onto an analog medium such as audiotape, and then digitizing the medium. If your computer has sound capability and you're able to connect it to a microphone or a CD or cassette player, you can use a sound recording program such as Windows Sound Recorder or Apple QuickTime. The next step, **edit**, requires cleaning up the sound quality, perhaps mixing it up with other sounds, or modifying it to make it fit into a specific time slot.

Real-time or Streaming Audio—*enables sound to play as it loads on the surfer's computer—the surfer doesn't have to wait until the entire file downloads before hearing it*

Step three, **process**, involves conversion, alteration, or modifications required to fit your sound into the format and tool that you plan to use for publishing. Finally, during the **publish** step, you'll need to select a tool to publish the sound file to your Web page. To accomplish all of these steps yourself, you'll use one or more of the many excellent software applications that have been designed just for that purpose, such as Macromedia's Director 8 Shockwave Studio (www.macromedia.com), Windows Media Encoder 7 (www.microsoft.com), RealNetworks RealProducer Plus (www.real.com), or Adobe's Premiere 6 (www.adobe.com). To begin with, however, you'll probably just download some sound files from the Internet and incorporate them into your own Web site.

If you want to add some background sound to your Web site so that it begins playing music automatically when the visitor brings up the page, it's probably a good idea to use either a WAVE or MIDI file, since most browsers are able to handle them. Make sure that you keep your sound file as small as possible, out of consideration for your visitors. Once you have selected the file—either you've created it yourself, or you've downloaded it from the Internet (be careful of copyright laws)—you can add it to your Web page through a number of methods, such as using a WYSIWYG Web design program, a tool such as Macromedia's Flash or RealAudio (which means that your visitors will require plug-ins), JavaScript, or HTML. Using HTML, there are several ways to include the sound file:

1. Use the <EMBED> tag, e.g.,

<EMBED SRC="filename.mid" >

The <EMBED> tag is supported by both Netscape and the later versions of Internet Explorer. It enables you to embed a file into your Web page so that it seems like an integrated part of that page, alleviating the need to open a separate window. A few extra steps you can follow to improve the effectiveness of your <EMBED> tag include:

AUTOSTART=TRUE advises the browser start playing the sound file automatically, as soon as the user surfs to your page.

LOOP=Value tells the browser how many times to play the sound file.

If you set the value to 2, it will play it twice; if you set it to 10, it will play it 10 times, etc.

A word of warning—secured sites such as corporate intranets may stop a Web page from loading if the Web page contains an embedded file using **<EMBED>**, because an embedded file can contain a virus. In addition, you should note that the file server on which the embedded object resides must MIME encode files of the type to be embedded, or the embed will fail.

2. Use the **<A HREF></A>** tags to link to the file, e.g.,

   <HREF= "woof.wav">Click here to hear my dog</A>

Between the link tags, specify the location of the sound file that you want to play. If the sound file is stored in the same folder as the Web page, specify just the filename. However, if it's in a subfolder, make sure you include the name of the subfolder, e.g., websound/woof.wav. When visitors click on the sound link, the sound will transfer to their computer and play.

3. Use the **<BGSOUND>** tag, e.g.,

   <BGSOUND SRC="Filename" LOOP=Times>

The **<BGSOUND>** tag specifies the sound file to be played automatically whenever someone has surfed to your site. In other words, it specifies the sound to be played in the background. You can tell the browser how many times to play the sound file from 1 through infinite. However, be careful of specifying "infinite" since this can really annoy your visitors.

"Times" represents the number of times you want to play the sound file, e.g.,

   <BGSOUND SRC="RocknRoll.mid" LOOP=2>

The <BGSOUND> tag is supported by Internet Explorer. Netscape users can experience the same effect through the use of the **<EMBED>** tag, e.g.,

<EMBED SRC="RocknRoll.mid" width=2 height=0 autostart=true loop=true>

Streaming multimedia files have become extremely popular because they start playing immediately, even while loading, rather requiring your visitors' browser to download the full file before any music or video starts. Adding streaming audio files to your Web site requires a few different steps than adding normal audio files. First of all, for visitors to play these files on your Web site, they will need to ensure that a player or plug-in such as RealPlayer or Microsoft's Windows Media Player has been installed on their computers. If these files do not exist, the browser or operating system will return a message to the user advising them that the file cannot be read. Help out your users by clearly identifying the type of file that will be downloaded, as well as a source for the necessary plug-in or player, or your advanced design may result in user frustration and fewer return visits. To create a streaming audio file, you will need to use an encoder, such as RealProducer Plus or Microsoft Windows Media Encoder 7, to convert audio input (e.g., from a microphone or file on your PC) to a streaming media file format. For more information and free downloads of software, visit the following sites:

MIME—*a multimedia encoding technique supported across multiple platforms as a way of distributing media file types*

Freebie Music
www.freebiemusic.com
Download free music software, as well as audio and real audio files. The site also includes some great games, e-zines, and much more. Definitely worth a visit.

Web Reference.Com
www.webreference.com
This site, an excellent all-rounder for general Web design information, has very good and specific information on how to incorporate multimedia into your Web site.

Yahoo's Audio Index
www.yahoo.com/Computers_and_Internet/Multimedia/Audio
This is a very good resource with numerous links to audio
sites and audio archives.

Archive Service of the School of Computing, Information Systems and Mathematics, South Bank University, London, U.K.
http://archive.museophile.sbu.ac.uk/audio
This site has an amazing number of links to sound repositories around the world. In addition to a list of sites for sound
files, there's an extensive list of audio software. Other items
of interest include newsgroups and online radio links.

Site Search Engines—Improving Site Navigation

As your Web site becomes larger and more complex, you will need to offer
your visitors help in finding what they need. Features such as site maps and
keyword searches (initially identified in Chapter 3, Good Design Basics, as
navigation tools) allow the visitor to speed up their search for specific
words or items within your Web site. In Chapter 9, Promoting Your Efforts,
you'll find information on some of the more popular search engines that
are external to your site; this section identifies some search engines you
can incorporate into your Web site, adding to the standard navigation set
up through nav bars and links.

If you're going to install a search engine, first check out a couple of trouble spots. Make sure that all your links successfully resolve and that your
<TITLE> tags include descriptive information, rather than just the name of
your company. Many search engines "build" their data by those two important functions. You can also work to enhance your search engines by "helping" your search engine improve its vocabulary. For example, make sure
your search engine maintains a log of searched-for items. Part of your
monthly search engine maintenance should include reviewing those logs
for "failed" searches. Perhaps some users type in "handphone" instead of

"handset" when searching for cellular handset information. You can then program your search engine to produce results for "handset" whenever the word "handphone" is used. Using the same rationale, you can also teach your search engine to correct common spelling errors or phrasing issues.

A well-planned site organization can also help the search engine to navigate your site, since many search engines use an "indexing robot" to search through your pages for the keywords, etc. You can also program into your search engine the directories you wish it to access; for example, if you don't want it to display all of your cgi scripts, you would not identify that directory as one to include in the search engine index. Usually, a standard file called the robots.txt file is located in the root directory of your site and prevents or restricts unauthorized indexing robots from accessing specific directories. Make sure your server or your Web host's server allows your search engine robot to index all directories to which you assign it.

Traditional search engines operate using keyword searches; however, Ask Jeeves helped lead the market into natural-language search engines. This handy format enables less technical users to search via a traditional question, such as "how many people can use the software?" In a keyword format, the user would enter search terms like "+license +users" to return documents identifying the number of users per license. Preference for the different search techniques vary greatly across the Web usage spectrum. Several Web sites offer keyword or natural-language search engine software that you can integrate into your own Web site:

Ask Jeeves Business Solutions
http://business.ask.com/software

Answers 5.0, a search engine software package that works like the popular www.ask.com site, enables a visitor to ask a question in everyday language. This eases the search process for users, rather than making them learn search criteria methods and identify the keywords necessary to produce the desired results. There's a demo and fact sheet available, as well as a form to complete for more information.

Inktomi Internet Search Solutions
www.inktomi.com/products/search

Inktomi provides a good selection of search products, including Search/Site. Its powerful search software incorporates easy-to-use natural-language inquiries. You can download a free trial version.

HotBot Help Tools
http://hotbot.lycos.com/help/tools

Through this helpful site, you can download HTML code to add to your Web site and incorporate some great HotBot search capability. There are several options from which to choose, including one for older browsers that don't support tables.

Google Customer Site Search
www.google.com/services

Google offers a selection of custom site search options that they host for you, including Custom WebSearch and Custom SiteSearch. The search tools enable users to ask Google to search your own site or the Web. Google hosts both options.

Flash Animation—Clutter or Art?

At some stage in your Web site development, you may determine your readiness to build in some advanced visual features—such as animation. Macromedia Flash, a powerful and versatile Web animation tool that made its debut in 1996, ranks highly as a star performer and shows no sign of diminishing in popularity. Although your learning curve may be a little steeper with Flash than with some other software packages, you'll find it well worth the effort to bone up on some of the basics of this jewel among Web visual design tools.

Once you start, you'll be hooked as you create "cool stuff" that adds another dimension to your Web site. Flash's functionality is based on vector graphics, rather than raster graphics, which means that your images will be sharper and cleaner, and, best of all, resizable without any loss of quality. There will be no pixel artifacts (lumps and bumps) that quite often plague the edges and backgrounds of images. Flash graphics will automatically resize whenever a visitor resizes the browser window—and still look great! In addition, Flash can build graphics animation files small enough to stream across most modem connections, which is another reason for its extreme popularity.

Flash can also create "tweening" animation, an exciting capability that will save you a lot of time and effort. To make use of the fantastic tweening process, start by building just the first and last frames of your graphic. Flash can then create the frames between the start and finish frames using the data you provided. It can even rotate the inside frames, if you wish.

Just as there are many devoted aficionados of Flash, there are also many equally passionate opponents, who would prefer to never see Flash used on a Web site. Therefore, before you get carried away and liberally use this exciting technology, take a look at the potential downsides of using Flash animation. First of all, be aware that moving images can be overpowering and will tend to dominate the vision and concentration of visitors who

Vector graphics—*graphics that use mathematical formulae to describe shapes, objects, and colors. Since the mathematical data is stored as plain text, a small amount of text can describe quite complex animation.*

Raster graphics—*bitmapped graphics comprised of colored pixels on the screen. Since it takes a lot of data to "describe" a pixel, files can become extremely large and take a long time to load.*

Tweening—*a mathematical technique in which an animation program generates extra frames between the key frames that the user has created. This gives smoother animation without the user having to draw every frame.*

view your Web site. It's hard to read text with half a dozen spinning logos and images taking on a life of their own. Opponents of Flash cite various concerns, such as the encouragement of the gratuitous use of animation (remember, we told you it was easy to get hooked on Flash!) by designers who feel that just because they <u>can</u> make it move, they <u>should</u> make it move. Don't let the quest for "eye candy" override good taste and common sense.

There is also an increase in the creation of Flash intros, the series of moving images, rather like a mini movie, that plays when you first go into a Web site. Intros can add value, when the content and form are carefully thought out, or can add pizzazz to an otherwise rather staid site. For examples, browse the Internet for sites that use Flash, such as OOPS (<u>www.oops.com</u>) and wolvesburrow (<u>www.wolvesburrow.com</u>). However, they can also be extremely irritating to your visitors, who must wait until the intro finishes before being able to see the "real" Web page. Bear in mind that not everyone has great connection speed and bandwidth. If you decide to add a Flash intro to your Web site, please add a "Skip Intro" button at the beginning of the intro. Your less patient visitors will thank you.

So, if the previous description of Flash has whetted your appetite, take a look at some of the many Flash resources available to you on the Web. Flash has quite a dedicated following, and there are some excellent Web sites out there. Here are just a few to get you started:

Flash Kit
<u>www.flashkit.com</u>

This great resource site for Flash developers includes relevant announcements, downloads, tutorials, and reviews. You can also look for the opportunity to show your Flash expertise (or see examples of others' quality efforts) through listed competitions, message boards, and chat forums.

The Flash Academy
<u>www.enetserve.com/tutorials</u>

This site starts off with an impressive Flash intro, complete with sound, and offers an extensive selection of tutorials, templates, source files, and resource links. This site accepts contributions from Flash fans and describes itself as a "free exchange of information, techniques, and ideas."

Flash 99% Bad
www.useit.com/alertbox/20001029.html
It's only fair to list a site that gives an alternative point of view to the use of Flash. This site describes some concerns and issues that are sometimes associated with Flash animation.

Video—Added Value or Not?

When you think of multimedia, you most likely think of both sound and video. After all, these are currently considered to be the pinnacle of multimedia special effects. Without video, you're probably going to feel that your Web site cannot hold its own with the truly greats! This is, of course, not necessarily true—you can have a perfectly splendid Web site without any audio or video. But, if you really want to impress yourself and feel extra clever at dinner parties, you should at least take a chance and investigate the opportunities video can provide for your site. A short video clip or live broadcast are both common examples of current video usage on the Web. For examples of video clips using a selection of formats, quality, speeds, and plug-ins, take a look at some Web sites that demonstrate effective use of video, such as:

ESPN
http://espn.go.com
Check out the sports videos. If you don't have RealPlayer, you can download it from here.

Multimakers

www.multimakers.com/video

This site allows you to view the same video clip at a variety of modem speeds, quality, and formats.

Windows Media

www.microsoft.com/windows/windowsmedia

This is the official Microsoft site describing Windows Media 8 for Video and Audio. There's a list of audio and video clips for you to try out.

Is video worth the effort needed to create quality results? Many Web designers will ask that same question at some stage of programming video for the Web. Although not particularly complicated, the process of capturing video, editing, and uploading is certainly involved enough to require some thought as to its value on your Web site. While it may be fun to create and add a video clip to your site, spare a thought for your visitors, who may have to wait for the file to download, or who may have to download a plug-in in order to view it at all. And always ask yourself if it's really necessary to have it. Does it add value to your Web page? Does it fit with the general theme of your site? If the answer to these questions is "yes," then go ahead and dip your toes into a highly exciting area of the Web.

And now for a little background about incorporating video into your Web site. Before a video signal can travel over the Internet, it must first convert from analog to digital format, then compress sufficiently to be

Analog vs. Digital—*analog format represents data in continuously variable physical quantities, in contrast to the digital representation of data in discrete units (the binary digits 1 and 0). Analog systems handle information represented by continuous change and flow, such as voltage or current. This is in contrast to digital information, which is either on or off.*

viewed in real time. Unfortunately, the higher the video's quality, the greater the file size. As a result, your Web visitors with slower Internet connection speeds and older computers will experience a less than satisfactory result. It can be disappointing to watch a video displayed as a series of jerky images, rather than the quality you're used to seeing on a television screen.

Like audio, there are essentially two types of video on the Web: downloaded video and streaming video. Streaming video is hot right now, particularly for broadcasting live events. It plays almost immediately—and after a few moments of buffering, viewers can begin watching the clip, whereas with downloaded video, viewers have to wait for the entire file to be downloaded before watching it. However, streaming video is generally of a lower quality than downloadable video, and most modem connections don't handle it well. It also falls prey to general Internet speeds and the power of the server that hosts it.

There are several downloadable video file formats available for the Web. The three principal formats have been around for a while but are still perfectly adequate in most cases:

- MOV (QuickTime)—Originally developed for the Macintosh, this format was created as a counterpart to Video for Windows. These file formats support true color, as well as multiple audio and video tracks.Macs have native support for QuickTime. Users of non-Mac systems will need to install a player or plug-in (available from Apple), unless their browser supplies native support.

- AVI (Audio-Visual Interleave)—This format, which contains both picture and sound, was developed for Video for Windows and is a standard for Windows 95 and up. It's now becoming less popular on the Web.

- MPEG (.mpg, .mp2, .mp3/Media Player)—The multimedia standards created by the Moving Pictures Experts Group include support for video, audio, and streaming information.Extremely popular for digital music, MPEGs retain excellent sound quality in highly compressed files, making them very sought after for

Web use. The latest versions of Netscape and Internet Explorer use Windows Media Player as the plug-in for playing MP3 files. This format creates smaller files with a high video quality and was developed as an international standard for use in CD-ROMs. You'll need a fast machine to be able to run them.

Although AVI and QuickTime are currently the most common files available for download, and can actually be previewed while they are downloading, they are not really streaming. The quality is generally good, however, and your visitors don't require any special software to view them. You should remember to try to keep downloadable video clips fairly small to minimize the waiting time for your visitors. To watch streaming video, your Web site visitor will almost always need to download and install a plug-in; see the end of this chapter for easy steps to downloading.

The main streaming video options are RealVideo, VivoActive, Videogram, VDOLive, Vxtreme, and NetShow. They all require the use of special software at some stage, e.g., development, serving, or viewing. For example, to develop streaming video, you need to use an encoder such as RealProducer. Some streaming video formats can be served from a normal Web server. Others require a dedicated media server, which will deliver all video formats more quickly and smoothly, but at some expense.

You'll also need a video capture card and video production software to produce video files. If your computer has a video capture card, you'll be able to connect it to a VCR, DVD player, or video camera. The majority of video capture cards come with the cables and software you need to record video. Make sure you have permission to use the video on your

Buffering—*the process of using a buffer, an area used to temporarily store data before delivering it at a rate different from that at which it was received*

True Color—*color system that uses at least 24 bits to represent over 16 million unique colors, more than enough since humans can only distinguish a few million colors*

Web page if you record footage that you didn't originally create. The technology used to produce streaming video is not difficult to use, since most video production software offers easy-to-use graphical interfaces. Macromedia's Shockwave, a plug-in with versions for both Netscape Navigator and Internet Explorer, can be downloaded at no cost to display files created with Macromedia's Director application. However, you will have to purchase Director or Adobe's Premier if you want to create the files yourself, but it is worth the cost. For more information on streaming media, take a look at:

Streaming Media World
www.streamingmediaworld.com
This site has a wealth of references, articles, tutorials, tools, and services. "What's Cool" contains plenty of showcased examples for you to view.

Always remember that high quality video files are BIG! Be kind to your fellow surfers and bear in mind that they'll need a fast connection to download a video file. The image size of your video clips will impact your site's performance, and the number of pixels in each video frame increases tremendously as you increase the size of the images. For example, a one-megabyte file that represents just a few seconds of video will take approximately 10-15 minutes to download over an Internet connection of 28.8 kbps. After all, you can't expect all of your viewers to have the latest and greatest fast connections.

You can add video to your Web page in the following ways:

1. Use the <A HREF > tag to set up a link to the video file, e.g.,

<A HREF="mymovie.mov"></A>

When your visitors click on the sound link, they will either be asked if they would like to download the file to their hard drive, or the browser will automatically download the video and play it in the viewer associated with the format connected with the file extension.

2. Use the <IMG> tag (Internet Explorer), e.g.,

> <IMG DYNSRC="Filename" LOOP=Times START=When>

"Filename" points to the video file you want to play, and "Times" specifies how many times the video should play. Note: if you specify "LOOP=INFINITE," the video will play continuously and your visitors will hate you! "When" tells the browser when to play the video. For example, START=FILEOPEN plays the video as soon as the user loads the Web page. START=MOUSEOVER will set the video to play whenever visitors move their mouse pointer over the video box.

Here's an example of a section of HTML code that plays a video clip for Internet Explorer:

```
<HTML>
        <HEAD>
                <TITLE> Squirrels in the Snow</TITLE>
        </HEAD>
                <BODY>
                Hi! To watch this movie—just move your
                mouse pointer over the video icon
                <P>
                <IMG DYNSRC= "squirrels.avi" LOOP=2
                START=MOUSEOVER>
                </BODY>
        </HTML>
```

3. Use the <EMBED> tag to embed the video file into your Web page, e.g.,

> <EMBED SRC="snowvideo.mpg" HEIGHT="200" WIDTH="200">
> </EMBED>

If you use <EMBED>, visitors will need a plug-in to view the video. For more information on specific video-related software and other useful tips, visit the following sites:

Apple's Final Cut Pro
www.apple.com/finalcutpro

Apple's official Web site has all the info on their video edit-ing and video management software. Final Cut Pro's real-time editing capabilities supports all professional video for-mats and boasts an impressive suite of animation, graphics, audio, and compression features.

Microsoft's NetShow
www.netshow.com

Gaining in popularity, Microsoft's NetShow is a streaming video solution that can be served from either an HTTP Web server or a dedicated media server. NetShow is free, and integrates well with Microsoft's Visual Basic if you want to combine text and images into synchronized events. CNET'S The Web (www.cnet.com) features a variety of NetShow products and product reviews.

Video on the Web
http://www.yourvideoontheweb.com

You can host your video from this site for free and share video clips with others. It's a great resource for video and tells you how to collect, capture, edit, and stream your video. You can get some great stuff here—tools, templates, and lots of advice on everything to do with video.

Five Steps to Downloading and Installing a Plug-In or Player

Since we've spent some time discussing an area of the Web that will quite likely require you at some time to download a plug-in to support your Web development, here are some basic instructions on how to proceed:

1. **Download the plug-in**—Check the "Save to Disk" option, when prompted, and click on the Browse button to locate the folder you created. Start the download into the selected folder. You should create a different subdirectory for each plug-in so they can more efficiently install. Make sure you know the directories to which your files download.

2. **Install the plug-in**—Once the plug-in file has finished downloading, close all programs running on your computer. Locate the plug-in file on your hard drive (The file will have a name like "plug-in.exe") and double-click on it. This will begin the installation process. Follow the instructions, restarting your computer, if necessary. If you encounter any problems with initial plug-in use, try restarting your computer before re-installing the plug-in.

3. **Tidy up**—Once you've installed the plug-in, delete the file(s) you downloaded into your temporary folder. This will free up disk space on your computer.

Popular Plug-Ins

To use many of the special effects features discussed in this chapter, you and your visitors will often be required to use plug-ins. The constantly changing Web environment continues to offer new plug-ins, seemingly on an hourly basis, so you may encounter some difficulties keeping on top of the latest information. However, these are currently some of the more popular audio and video plug-ins used on the Web:

+ Shockwave: <u>www.macromedia.com</u> Shockwave supports interactive multimedia, streaming audio, and graphics.

+ RealPlayer: <u>www.real.com</u> RealPlayer, from Networks, plays streaming audio, video, multimedia, and animation.

+ Flash Player: <u>www.macromedia.com</u> Flash Player displays 3-D graphics, animation, and arcade games. It's one of the most popular plug-ins and works with Windows and Linux.

- QuickTime: **www.apple.com** QuickTime supports streaming audio and video, 3-D 360% navigation.

- Windows Media Player 7: **www.microsoft.com** One of the more popular players, Microsoft's Media Player 7 can be downloaded free for Windows 98 and 2000.

8

Troubleshooting Your Efforts

Now that you've put in all the hard work of building an excellent Web site, it's important to ensure that your efforts are effective and that you don't fall at the final hurdle. Testing is an essential step in deploying an effective and worthwhile Web site, and unfortunately one that many Web designers don't take seriously. In the early days of the Internet, most people were fairly tolerant of the first, often imperfect efforts of those daring Web design pioneers. Nowadays, however, there is a more sophisticated surfer, a greater commercial need for the Internet, and more and more highly skilled Web designers. People are generally no longer as tolerant of typos, bad links, poor design, and bad content—nor should they be. But don't let any of this put you off making your contribution to the Internet. Just be aware of the things that can go wrong with your Web site, and take the time to test for, and find, any problems before your visitors do. You'll be very glad you did.

So, what exactly can go wrong and what should you be looking for? Obvious errors include incorrect spelling, poor punctuation, and bad grammar that make your Web site appear unprofessional, as they would in paper format. Prepare your Web documents with the same care as traditional documents—take the extra time to make those easy fixes. Look out for a problem common to many Web sites—broken links. Create a successful file and directory structure, and know it well by keeping all of your files and graphics nice and orderly in their Web site directories. Beware of moving or renaming files once you've linked them, or you may find that the links no longer work. Luckily, programs like the WYSIWYG editors described in Chapter 2, The Tools You'll Need, will automatically change links within the files and prepare them to upload to your site. Incorrect code or programming not supported by a specific browser represents another frequent mistake. For example, not all tags are equally well supported by the main browsers, and some browsers won't respond well if you fail to include a closing tag in your code.

Take care that you establish and follow good design standards. Poor design that makes it impossible for visitors to see your content clearly or navigate the site readily can be easily avoided (see Chapter 3, Good Design Basics). Web page size (height and width) can also cause problems. With the varying sizes of monitor screens, oversize Web pages with large graphics may become almost unreadable on smaller monitors. On a Windows PC, the browser will default to displaying the page to the full width of the screen. On a Mac, it may be narrower so as not to hide the icons on the right-hand side of the screen. Also, test your Web site for download times. Most users hate waiting for slow-loading pages that contain large, bulky graphics and may give up and leave before your site finishes loading. If your testing reveals this as a problem, try keeping the top portion of the Web page relatively free of large graphics, since this will give the impression of a faster download. Chapter 6, Your Best Image, provides some great tips for "image optimization," which enables you to create quicker loading graphics that meet high quality standards. Alternatively, you can use "progressive rendering" or "interlacing," during which the entire image appears in low resolution almost immediately, then gradually becomes sharper. Many people prefer this than to wait for the image to appear little by little.

You should manually test the above trouble spots, as well as other factors, unless you have an extremely large and complex site, or time presents a barrier. In those cases, you can select from the many excellent Web testing software packages available for purchase and download from the Internet. Alternatively, special Web sites can test your site for you and quickly report the results by email or on your screen. Whichever method you choose for Web testing, here's a quick checklist that you can use to cover the basics and get you started:

Quick Test Checklist—Ten Potential Problem Areas

1.	Spelling, Grammar and Punctuation	Is everything spelled correctly? Are there any noticeable grammatical errors?
2.	Links	Are there any broken links, or links that go to the wrong place? Are you missing any links?
3.	HTML	Does the HTML coding contain any errors, or is anything not supported by the different versions of the browsers?
4.	Browser Compatibility	What does the Web site look like in the various main browsers?
5.	Navigation	Can the visitor easily move around the site and get back to the home page easily? Do you have a site map, table of contents, menu bar, etc.?
6.	User Interface and Accessibility	Is the Web site easy to use—do all buttons and other clickable objects work correctly? Is there a consistency of design from one page to another? Do all similar objects have the same "look and feel"? Have you taken into consideration accessibility for the disabled?
7.	Load and Performance	How fast do the pages/graphics appear, etc.? How many visitors do you expect to access your site at a time? How well does it perform at different connection speeds?
8.	Content	Is the content correct and up-to-date? Do you continually maintain it so that it's not static?
9.	Contact Info.— Feedback Forms and Email	Do all your contact features work correctly? For example, does the email work? Does the feedback form accept data and validate data effectively?
10.	Screen Size and Resolution	From 640 x 480 pixels to 1024 x 768: How does your page look on the most common PC screen sizes and in the different browsers?

Each of the preceding tests can be accomplished through the use of online test resources (see the *Online Resources for Testing and Reviewing Your Web Site* section later in this chapter).

Browser Basics

Never fall into the trap of thinking that just because your Web site displays well in your favorite version of a browser, it will be fine in all the other browsers and versions. Browsers often exhibit idiosyncrasies that cause Web pages to be displayed differently; whereas some browsers will forgive your minor HTML coding errors, others will not. And, no, please don't just slap a "Best viewed with xxxx browser" label on your Web site and hope for the best! You'll quickly drive away users who do not surf with that browser, and they most likely will not return.

Once you're sure that your Web site contains no coding or link errors, you'll want to make sure that the majority of visitors see it at its best. A good rule of thumb is to check how your page looks in the current version of each of the main browsers, as well as in the two preceding versions of each. At present, although there are many different browsers on the market, the major players are Netscape and Internet Explorer. Others currently in use include Mosaic, Opera, and text-based browsers such as Lynx. As a Microsoft product, Internet Explorer is included on most Windows computers, although it and the aforementioned browsers can be downloaded from the Internet. Keep in mind that at least 80% of surfers use some version of either Internet Explorer or Netscape.

Web Browser—*a computer program used to view and download Web pages on the World Wide Web. A browser is designed to read HTML code and display the Web page on your computer screen*

If you experience repeated, frustrating problems with the look of the text in various browsers, bear in mind that text is comprised of three elements: size, style, and font. Some of the more unusual fonts may cause display problems in some browsers or versions. Additionally, in order for a font to display on a user's computer, that font must reside on the user's computer, or the browser will default to the standard font. Likewise, font size that is too small or too big may cause problems for the visitor. For example, if you use a font size that is too small, many people over 40 will have difficulty reading the text. If the font size is too large, it may cause layout problems and could be irritating to the visitor. You don't want to drive people away from your site because of something easily fixable, such as font size. Most browsers like a font size of 12 or 14 for standard text, and many usability experts recommend using at least a 10-point font size.

Graphics can also be a troublesome area for browsers. Not all graphic file formats are supported by all browsers, and visitors may require a plug-in or external software package to view them. When you're testing your site, take a good look at all the advanced, higher-bandwidth elements you've added—graphics, animations, Flash, Java applets, sound, and video—and make absolutely sure that you really need those elements, and that visitors can easily download any required plug-ins. When you're testing your site, remember that, just as colors might not look the same in different browsers, not all fonts and HTML code are supported, nor will all graphics display the same across the different versions of the various browsers. In fact, some coding results, fonts, and graphics may not display at all. Additionally, if you've used frames, check that you've added the **<NOFRAMES></NOFRAMES>** tags. All graphics should include alternative text tags (just add **ALT="text"** into your graphics tags). If your graphics do not display for any reason (especially to visitors who have turned off graphics on their browsers), the alternative text will provide some contextual reference for those graphics.

In addition, you can give yourself and your Web site an excellent head start if you take Web standards seriously. The World Wide Web Consortium (W3C), founded in 1994, develops common technology standards for the Web. Its members include technology content providers, corporate users, research laboratories, standards bodies, and governments. Unfortunately, although browser software designers know of the importance of the standards established by W3C, some differences invariably find their way into the software, partly due to the competitive natures of the big browsers. A growing number of Web developers and designers take these standards seriously, so take a moment to check out the following sites that offer validation tools and services:

W3C HTML Validation Service
http://validator.w3.org

Feeling brave? Check out your Web site against the W3C standards. Their free HTML Validation Service will take any URL you supply and run its HTML code against a long list of coding standards. The report is very nice and cleanly laid out, and they'll let you know if your site makes the grade with regard to validating as HTML 4.0—the latest version as this book went to press.

HTML Tidy
www.w3.org/People/Raggett/tidy

HTML Tidy, a free utility, checks your HTML code and corrects it or alerts you to take another look at the flagged "warning" or "error." You can even specify the layout style you want Tidy to use when correcting your code. Versions of Tidy support various operating platforms including DOS, UNIX, and Windows. Tidy can be downloaded free from this site.

Of course, you've followed all the advice for good Web design and taken steps to make your site compatible with the main browsers, but there's always the possibility that some favorite feature or piece of fancy formatting is still causing browser problems. Too many differences exist between the various combinations of platforms, browsers, and versions to list them all here, but you can find some excellent browser compatibility charts on such sites as WebReview: www.webreview.com/browsers. Should this be the case, be prepared to make some tough decisions—should you compromise your artistic creativity and eliminate any "problems" or leave them in and sacrifice part of your audience? We'll leave this decision up to you! While definitely not comprehensive, here are a few of the main differences (as this book went to press) between the most popular browsers for which you should be on the lookout:

Netscape (http://home.netscape.com)

One of the two most popular browsers, Netscape is less forgiving of HTML errors than is Internet Explorer (IE), so be sure your tags are well checked! In addition, it does not support the scrolling text **<MARQUEE>** tag or floating frames **<IFRAME>**. Other common tags and attributes that Netscape does not support include the table attributes for **background**, **bordercolorlight**, **bordercolordark**, frame and rules; and color of horizontal rule **<HR COLOR = color>**. Netscape does support the **<BLINK>** tag, which causes your text to blink, and **<MULTICOL>**, which creates multicolumn (newspaper style) text columns. Additionally, at the time of this book, Netscape 6 provided the most comprehensive support for Cascading Style Sheets, as well as excellent support for Java, XML, tables, frames, and JavaScript.

Internet Explorer (www.microsoft.com/windows/ie)

One of the most widely used browsers on the market, Internet Explorer (IE), is bundled with the Microsoft Windows operating system. At the time of this book's publication, IE's newest version, 5.5, provided good support for Cascading Style Sheets, as well as excellent support for Java, XML, tables, frames, and JavaScript. HTML tags that are not supported by IE include **<SPACER>**, **<BLINK>**, and **<MULTICOL>**. IE cannot utilize two handy Netscape features: Print Preview and the ability to right-click on graphics to view them individually. IE supports the **<BGSOUND>** tag that enables you to add background sound to your Web site, but Netscape does not.

Opera (www.opera.com)

Opera prides itself on being the World's fastest browser, although it is smaller in both program size and popularity than Netscape and IE. Consequently, Opera provides only partial support for Java, CSS, JavaScript, and XML. The first browser to seriously address accessibility issues, Opera includes many features like hearing prompts and zoom capabilities for the visually impaired and keyboard shortcuts for the mobility impaired. While this up-and-coming browser still has a far way to travel before catching up with the big browsers, many users are moving to adopt this program.

Lynx (lynx.browser.org)

Lynx is one of the most common text browsers currently used on the Web and is popular with users of both Unix and VMS platforms who are using cursor-addressable, character-cell terminals or emulations. Lynx 2.8.3 (the most recent version at the time of this book) runs on Windows 95/98/NT, DOS 386 or higher, and OS2 EMX. Lynx is also particularly well used among the visually impaired since it works well with screen readers and Braille displays. To see what your Web site would look like to the Lynx text browser, try it out on a Lynx Viewer from www.delone.com/web. Just enter your URL and see it displayed in the viewer.

Some online Web sites that provide more information on browser compatibility include the following:

Web Monkey—Sizing up the Browsers
www.webmonkey.com

It's definitely worth taking a look at this site with regard to best uses of browser testing. Helpful hints cover dimensions and resolutions that produce the most effective results on different browsers. A great browser chart also shows the different features supported by the various browsers and browser versions.

Browser News Resources
http://browserwatch.internet.com

With breaking news on the latest and greatest in browsers, this site will keep you informed about browsers, plug-ins, and ActiveX controls. BrowserWatch-Browser Blvd provides up-to-the-minute information on platforms that are supported, or planned to be supported, by a variety of browsers. If you ever doubted the existence of browsers other than IE and Netscape, browse through their A-Z list of "All the browsers we know about."

Also, see the *Online Resources for Testing and Reviewing Your Web Site* section of this chapter.

How Friendly is Your Web Site?— Testing Usability

How easy is it for your Web visitors to find what they came for? How comfortable do they feel? Can they actually read and comprehend the text, efficiently move around your site, and return Home with only one click? In preceding chapters, we've covered the essentials of good design, navigation, "look and feel," browser compatibility, and many other features and issues that constitute a "user-friendly" site. When you're in testing mode, you should be re-thinking these issues and checking them off one by one. We've included a basic usability checklist for you at the end of this section.

Usability is currently one of the hot topics on the Internet and has become something of a "must have" buzzword when it comes to discussions on Web design. Chapter 3, Good Design Basics, covers the essentials of usability design. When it comes to testing, keep three things firmly in mind:

1. Who are the users?

2. What do they want/need to do on this site?

3. Is it easy for them to accomplish this? For example, have you ever

visited one of the older travel-booking sites where all you really wanted to know was the cheapest fare to Paris on such and such a date? But to try to get this information involved a lengthy and painstaking journey through a myriad of screens that eventually led you back to where you started—and you STILL didn't know the cheapest fare! Try to avoid this on your Web site by spending a good chunk of your testing time on this issue.

If you're serious about testing your Web site for usability, think about conducting some monitored user testing. Let friends and co-workers try out your site while you watch, and take note of how your "testers" navigate. Also observe what they stop to look at, their positive and negative reactions to different site elements and features, and ask for their feedback on their overall site experience. You should always work with test users, since by now you may be so familiar with your site that you will experience difficulty looking at it with fresh perspective. If you're lucky enough to have a whole team of user-testers, assign each tester to a specific usability area. For example, one tester can concentrate on navigation, one on speed, another on consistency and layout, and so on. If you have time, switch roles to get different perspectives. Of course, if you're doing all the testing yourself, you'll have to wear all the hats and do everything, but try to concentrate on one area at a time in order to stay focused.

Usability Test Checklist

1.	Navigation	• Is there a site map?—this is a good idea if your site has more than 30 pages. Do all the links in the site map work? Can visitors easily find the site map from any page?
		• Do all the clickable icons, buttons, graphics, and text links work correctly? (You may want to do this test often, particularly if you have any links to external sites.) Can the visitor get back to the Home page easily and from any other page?
		• Do you have a navigation bar or column on each page? Do all the tabs or text links work correctly?

♦ Do all the navigation aids on your pages have a consistent look and behavior to them?

♦ If you have a search engine on your site, does it work well? Try entering some keywords and see what happens.

♦ Do your hyperlinks have clear descriptive text so that the user has an idea to where he/she will be jumping? Don't just label them "Click Here."

♦ Are META tags provided on key pages?

2.	Consistency	♦ Are all your logos, navigation bars, and navigation buttons in the same place on each page?
		♦ Do all the user interface elements on your Web site behave in the same way? For example, if you click on a button, does it do the same thing as a similar button on another page?
		♦ Do all your error messages, link text, and help text have the same wording where appropriate?
		♦ Do all your pages have a proper title that is consistent with their content?
3.	Color	♦ Have you used the same color combinations for each page?
		♦ Have you used color appropriately?
		♦ Does the color look similar in all browsers?
		♦ Can you see the text clearly against the color of the background?
		♦ Are links (visited, non-visited, active) using appropriate colors?
4.	Text	♦ Are you using all caps? (You shouldn't be—it's harder to read.)
		♦ Is the text an appropriate size and easy to read?
		♦ Is the text appropriately justified so that it's easy on the eye? (Most people prefer to read the majority of text as left-justified.)
5.	Layout	♦ Do you have sufficient white space, or does your page look cluttered and "busy"? Try to keep it to no more than 50%.

		◆ Are all elements of the screen grouped appropriately? For example, logically connected text and graphics should be grouped together. Use boxes or white space to separate groups.
6	Windows	◆ If a visitor opens an additional window, is it centered? (It should be.)
		◆ If a separate browser window is opened, is your site's navigation bar or logo still retained at the top, or does the visitor get lost in the external site?
7	Speed	◆ How long does your Web site take to "come up" in various browsers and at different connection speeds? If you have many large graphics, the time to download may be excessive. Remember, surfers are not known for their patience.

Some good sites to take a look at with regard to usability issues are:

Usable Web
http://www.usableweb.com

With over 1,200 links about usability, this site provides a plethora of helpful information in determining your site's effectiveness. Browse through links about information architecture, human factors, user interface issues, and usable design specific to the World Wide Web. You'll also locate the hottest usability books, related events and organizations, and links to other good usability-related sites, through which you can learn more about the topic. It's well worth a visit.

SURL Optimal Web Design
http://wsupsy.psy.twsu.edu/optimalweb

This is the Web site of the Software Usability Research Laboratory (SURL) at Wichita University. The site has some excellent information on many of the most common usability issues, including readability, color, navigation, and the organization of information. It also provides a hyper-linked resource list that will prove handy in your search for more usability support.

Content Credos to Live By

Unfortunately, content blunders occur, and if you're designing a Web page for an organization, they can result in embarrassment or carry legal implications for the company or for you. Minimize your liability by correctly following the review processes established for your company, which might include a variety of steps involving your peers, department heads, legal counsel, or even the CEO, depending on the content and liability involved. Whether you function as a Web designer, programmer, or content manager, you should protect yourself by keeping an email or paper trail, or combination of both, until after the pages in question have been launched and you are positive no further questions will arise.

Generally, small companies (and of course, enthusiastic individuals creating personal sites) may depend on one person to manage the coding and content of a site (whether alone or through consultants). In this case, concentrated responsibility will rest on that person's shoulders, as the others may not have the time to review content for a Web site. However, do your best to get someone else to review your text for you, whether spouse, friend, co-worker, or peer.

In a larger company, the bureaucracy of Web content review may seem overwhelming, by contrast. In that case, you may encounter several layers of review, though three to four sets of eyes should prove more than sufficient for your needs. Work with the peers in your department for quick reviews and ensure that you engage the owner of the content, especially if they reside in a different department and you support them in a consultant-type relationship. Additional reviewers whose time and efforts desire the utmost respect and deference include members of your legal and senior staff—ultimately, they know best the implications of your work. Do not rely on them to proofread your work but, rather, to ensure that your information effectively satisfies the mission, goals, and trends of the company.

Standard Mistakes to Diligently Avoid

Spelling. Plane too sea, mispellings dew knot luke gud four ewe.
Unfortunately, half of the wrong words above would not be caught through a spellchecker, and any dictionary would locate them as correct words. However, you probably noticed that the homonyms (words that are pronounced alike but are different in meaning and sometimes spelling) really don't generate the professional image you seek. These and typographical mistakes occur more frequently than you could imagine, and although they might seem silly or harmless, they degrade the professional quality of the material you seek to present. Spend a little bit of time and money on any of the following books, which provide excellent references for frequent questions on style and grammar:

- *Woe is I*, by Patricia O'Conner

- *Write Right! A Desktop Digest of Punctuation, Grammar, and Style*, by Jan Venolia

- *The Elements of Style*, by William Strunk Jr., E. B. White, Charles Osgood, Roger Angell

Although spellcheck tools in word processing and Web design programs, and those on the automated testing sites, might catch some of the major mistakes, they can also cause other mistakes. For example, if you quickly proceed through spellcheck for an online event program you are preparing for your Web site, the spellchecker may return some very odd results to replace unusual and unrecognized last names. Therefore, though you can use spellcheck, you should always do one last review of a document in the final development stage before information is posted live.

If you are acting as the proofreader or editor for someone else's work, you need to properly prepare yourself before beginning. First, ensure that you have a clear understanding of the role which that person intends you to play. If a proofreader, you should expect to conduct a quick scan of the text provided, looking for misspellings, comma use, and the standard mistakes listed below. If you need to view the text with an editor's eye, not only will you consider all of the common elements a proofreader would catch, you should also consider ways in which the writing could be improved, especially the flow of ideas. The biggest challenge of editing requires you not to change the basic style of the writer, but to capture bad style and mistakes that could prove embarrassing. Make sure you and the person whose material you are reviewing have a clear, common understanding of editing/proofreading marks that you will use. You can find a good list of standard proofreading marks at **www.prenhall.com/author_guide/proofing.html** and **www.alaska.net/~mjedit/proof.html**. Other excellent sites to help you with your editing or proofreading challenges include:

EEI Communications
www.eeicommunications.com
Although a commercial site that first and foremost seeks to highlight EEI's services, this site includes buried gems that you won't want to miss. Take advantage of its search engine to look for proofreading, for example, and you'll gain access to a list of proofreading marks and excellent articles from *The Editorial Eye*, a very helpful periodical.

The University of Victoria's Hypertext Writer's Guide
www.clearcf.uvic.ca/writersguide
This extremely useful Web site provides information on everything from documentation and word use to logic and general grammar use. Although designed for English students, it presents one of the most comprehensive resources for editors and proofreaders.

Whether working with your own or someone else's work, keep your eyes open for little typographical errors, as well, and ensure that the content on your site follows consistent standards, whether established through a company style guide or in your head. Watch for frequently repeated words, as well, and if you do not own a dictionary and thesaurus, get one of each now, so that you can quickly and easily take care of problems. You should not end up with three different versions of a product name, whether through spelling or capitalization errors, which may decrease the brand value of that particular product name. By setting standards as described in Chapter 4, Content to Keep 'Em Coming Back, through a style guide, no matter how formal or informal, you protect both yourself and your content.

Frequent Typographical and Spelling Errors

accept/except	decisions	possessives/apostrophe use
achieve	development	proceed/precede
affect/effect	disinterested/uninterested	questionnaire
all right/alright	etc.	success
a lot/alot/allot	good/well	than/then
beside/besides	i before e, except after c	job titles/departments
can/may	its/it's	to/too
comma and semicolon use	lie/lay	whether or not
committee	like/as	who/which/that/whom
contractions	neither/nor	whose or who's

Online Resources for Testing and Reviewing Your Web Site

Fortunately, there are plenty of excellent online resources available to test your Web site in some of the trickier areas—or even in some of the easy areas if you need to save some time. For example, running an HTML validation tool against your code will highlight any errors that may exist, such as missing closing tags, tags that are incompatible with standard browsers, missing attributes, and so on. In addition to downloading and/or purchasing test software that you can use to test your Web site, there are Web sites that will do the testing for you—all you need to do is provide them with an email address and the URL of the Web site you would like tested.

Tests provided by these online services include HTML validation, link checkers, and spellcheckers, as well as evaluation of download speed, graphic size, browser compatibility, and handicap accessibility. There's definitely plenty for everyone and certainly no excuse for not having a well-tested and professional Web site. Great online testing tools and resources include, but are definitely not limited to:

WebKing
www.thewwebking.com

WebKing, a comprehensive Web testing and test management tool, perhaps more appropriate to organizations than individuals, tests the functionality, construction, design, and content of your Web site and builds test scripts to emulate virtual users for load and stress testing. In addition to its testing features, WebKing can organize your multi-tiered Web site development process, including ways to organize and access files from multiple sources and machines, into a common source code repository. The free demo will even help you assemble, maintain, and reproduce staging areas and Web sites.

HTML Toolbox
www.netmechanic.com

NetMechanic's HTML Toolbox 2.0 offers an excellent tool for testing your Web site. After you enter the URL you want to check, detailed results will be quickly delivered to your email address. The free trial will check up to five pages in your designated site and conducts the following HTML tests: link check, spellcheck, HTML check, and speed (of downloading graphics and pages) check. The site clearly identifies and explains the browser incompatibilities in your HTML code. If you like the results, you can purchase an annual service contract.

Browser Photo
www.netmechanic.com

Also by NetMechanic, this handy tool helps you to ensure that your Web site can be seen equally well by all browsers. Browser Photo facilitates this task by taking screen shots of each page as it would be seen in 14 different browser/computer combinations, allowing you to see the results and fix your Web site, if necessary.

Lift Online
www.usable.net

This online testing service will test your site for HTML coding errors as well as uncover problems with usability and accessibility. You can try out the service for free by entering a URL and valid email address. The results will be sent to you in approximately five days.

Handicap Accessibility and Why It's Important

To truly ensure that your Web site meets usability standards, you should consider a number of issues regarding accessibility to people with disabilities. As the Internet becomes more popular and worldwide, it will become increasingly important as an integral part of our everyday lives in the future. It's important, therefore, to ensure that everyone can use and benefit from this amazing resource, and that means being aware of potential problems and difficulties that some Internet users may have.

Accessibility—*a measure of how easy it is to access, read, and understand the content of a Web site*

For example, visually impaired people using a screen reader (a tool that reads the text aloud or translates into Braille) may have problems if the graphics, JavaScript, video clips, and even links on your

Web site do not use alternative text equivalents, also known as <ALT>tags. Make sure you name these elements in a descriptive manner so that your visitors with disabilities will enjoy a successful time with your site. GUIs pose barriers for the visually impaired, who generally use a text browser, such as Lynx, so make sure that your HTML code is text-browser friendly. People who cannot differentiate colors well or who are color-blind may also encounter difficulties with Web sites. Colorfield's Insight (www.colorfield.com/insight.html) is worth checking out when designing your site; it can help Web designers model and predict the image legibility for color-deficient viewers.

Internet accessibility is definitely improving for the disabled. The World Wide Web Consortium (W3C) has developed a set of guidelines that support their Web Accessibility Initiative (WAI) (*Web Content Accessibility Guidelines 2.0, W3C Working Draft, 25 January 2001*). Tim Berners-Lee, W3C director, is quoted on the W3C Web site as follows: "The power of the Web is in its universality. Access by everyone regardless of disability is an essential aspect."

Quick Checklist for Testing for Handicap Accessibility

1.	Links	Instead of "click here," make the text descriptive (for when it's read aloud by a screen reader).
2.	Video	Provide a brief but meaningful alternative text description of the video.
3.	Frames	Include <NOFRAMES> tags and meaningful text.
4.	Graphics and Java Applets	Use <ALT> tags and meaningful alternative text—make your code text-browser friendly.
5.	Colors	Make sure they display well in multiple browsers. Note any colors with which the visually impaired may have problems.

Some great sites to take a look at with regard to designing accessible Web sites include:

Web Page Accessibility Self-Evaluation Test 2.0
www.psc-cfp.gc.ca

The Public Service Commission of Canada hosts this accessibility test that you can try for yourself online, or download as a file to peruse at your leisure. This is quite comprehensive and definitely worth taking a look at. There are 27 topic areas such as graphic links, frames, browser-specific HTML, and design consistency—with several questions in each section.

The Productivity Works of Trenton
www.prodworks.com

WebHEARit enables any Internet Explorer application to "speak" a Web page's content and navigational links. It improves on existing programs that read computer screens aloud. The software enables blind and other disabled users to browse through the headings and highlighted hyperlinks on a Web page, finding what they want and jumping from page to page like a sighted person.

Bobby
www.cast.org/bobby

The Center for Applied Special Technology (CAST) offers Bobby, a free public service. This helpful Web-based tool analyzes your Web pages against the WAI Guidelines 1999 for accessibility to people with disabilities. Just submit the URL of the page in question, and Bobby will display a report that indicates any errors found in accessibility or browser compatibility. The resulting report can be helpful for any testing purposes, since it includes trouble points such as alternative text, color contrast issues, navigation, download time, etc. Bobby only tests one page at a time, but if you'd like to test your entire site, there's a downloadable version. Once you pass all of the Priority 1 checkpoints established by the WAI, and achieve "Bobby-approved" status, you may display a special graphic on your Web site indicating your support of handicapped accessibility.

WebABLE
www.webable.com

This very useful site provides information helpful to your research on disabilities and the Web. It includes an extensive list of links to accessibility-related seminars and workshops, books, Web sites, design tips, and more.

9

Promoting Your Efforts

Web site promotion offers many creative challenges to those responsible for ensuring that their Web site receives the full attention that it deserves. When corporations, small businesses, consultants, and individuals reach this stage in the Web site game, they or other teams have likely spent days, weeks, or even months, with design, content creation, and programming invested to create the best Web site ever. Especially when Web sites have been designed for a profit-oriented company, people will expect results, in the form of visitors, which support the purpose of the site.

Now, you need to let people know where to find your site in the complex infrastructure of the World Wide Web. As people push forward with their site and thoroughly integrate themselves with the excitement of the online marketplace, they frequently forget the traditional, print methods through which you can feature your Web site. As you browse through the online promotional methods, such as search engine submission, online newsletters, meta tags, etc., don't forget to utilize your print alternatives, as briefly described in the domain names section below. Although this chapter will cover some of the different types of tools and resources, the following books offer a more in-depth look into the variety of resources available to help maximize your site's reach:

- *Marketing on the Internet,* by Jan Zimmerman

- *Internet Marketing for Dummies,* by Frank Catalano, Bud E. Smith

Domain Name Registrations

Domain name registrations represent one of the quickest and easiest methods for maximizing your marketing and advertising reach. They help you create or enforce your branding strategies on the Internet. Your domain name will appear at the top of your Web site, in links, in search engines, and on every email you send. It establishes your digital address and conveys the image of your organization or project. People have named their companies after domain names they acquired, to cement the branding and ensure that Internet users can immediately locate their Web site. At the time of this book's composition, over 30 million domains had been registered. After you select your domain name, maximize its potential by advertising it everywhere—in print (on your business cards, company letterhead, etc.); in electronic format (in email signatures, on your Web site, in online newsletters), and anywhere else you can place it.

The price for creating a standard domain name can vary, depending on whether you choose to register a new name, through a registrar like Network Solutions (www.networksolutions.com), or if you decide to purchase a pre-registered domain name, through a site like Great Domains (www.greatdomains.com). If you opt to create a new domain name, you may encounter some frustration when registering a name in one of the traditional (.com, .net, and .org) top-level domains, but don't give up too easily...plenty of names exist; you just need to creatively choose which one works best for you.

On the other hand, if you decide to purchase a pre-registered domain name, you may trade in creative thinking for a high-priced name that immediately satisfies your needs. Some of the higher-priced domain name transactions have included the following:

- Business.com $7,500,000

- AsSeenonTV.com $5,000,000

- Loans.com $3,000,000

- ◆ Beauty.cc $1,000,000

- ◆ Drugs.com $823,456

Of course, not all previously registered domain names generate such high price tags. Some may cost nothing more than the registration fees. Just choose wisely and make sure you promote your domain name as thoroughly and effectively as possible.

Whichever method you choose, consider purchasing more than one domain name. First, you can quickly and easily assign more than one domain name to your site. Do you offer particular products or services with unique names? Add those to your list of domain names to protect their online identity. Also, especially if you have great plans for your Web site, register more than just the .com domain name. Protect your identity by also registering .net, .org, .cc, .tv, and the new domain names that will soon be offered (see <u>www.icann.org</u> for more information). You can also register your domain name in different languages and in different countries, extending your reach and protecting your valuable online branding tool from people and companies that might imitate it and lower your domain's value.

Search Engines

Now that your Web site stands prepared for the world to discover, which tools will you choose to best ensure that people can find your site? Web crawlers and search engines present an optimal resource for announcing your site to the world. The first, Web crawlers/spiders, require some initial preparation on your part with meta tags. Web crawlers "crawl" through the Internet, looking at all of the existing pages, capturing the meta tag information therein, and placing it into a directory of Web sites through which users can search for information. Although you need only prepare the initial meta tags through

> **Meta tags**—*special tags that appear in the <head> of your HTML documents, to identify for search engines the content and purpose of your site*

which they can search, you can increase and improve your chances of being noticed, by ensuring that you know the hot keywords used for Internet searches. Search engines sometimes use Web crawlers to help identify pages, but also require that Web site owners submit their sites. Tools and strategies for successful submissions appear later in this section.

Top Ten Search Engines

1. AltaVista: www.altavista.com
2. AOL Search: http://search.aol.com
3. Ask Jeeves: www.askjeeves.com
4. Excite: www.excite.com
5. Google: www.google.com
6. Hotbot: www.hotbot.com
7. Lycos: www.lycos.com
8. Netscape: www.netscape.com
9. Northern Light: www.northernlight.com
10. Yahoo: www.yahoo.com

Meta Tags

Each of the search engines uses your meta tag information slightly differently. Some search engines will ignore meta tags, instead reporting the information in your <TITLE> tag or the first 250 characters in your Web site; others, however, rely heavily on your keywords and description, and can determine whether your site flies or falls in search engines. Meta tags occur in an attribute/value pair consisting of two main attribute types, HTTP-EQUIV and NAME, and a value, CONTENT:

<META HTTP-EQUIV="insert type of HTTP-EQUIV"
CONTENT="identify value">

<META NAME="insert NAME here" CONTENT="identify value">

The CONTENT attribute specifies a certain value to the property mentioned in the NAME or HTTP-EQUIV attribute. Fortunately, some excellent Web sites help identify the ins and outs of working with meta tags:

Vancouver Dictionary of Meta Tags
http://vancouver-webpages.com/META
This comprehensive resource for meta tag information defines the different attributes and pairs used in meta tags in a very easy-to-follow format. The site also offers a great FAQ that will enhance your understanding of this sometimes complex topic.

Spider Food's Meta Tag Tutorial
http://spider-food.net/meta-tags.html
This tutorial highlights the triumphs and tragedies associated with meta tag work. It highlights the use of keywords and features a wide variety of useful information about optimizing your site for search engines.

HTTP-EQUIV

HTTP-EQUIV tags help control the actions of the browsers in relation to the Web documents and can play an equivalent role to HTTP headers. Because some servers actually generate HTTP headers from the HTTP-EQUIV tag, you should not invent new HTTP-EQUIV tags unless you understand the HTTP specification process; use NAME tags instead. They usually occur in conjunction with the CONTENT tag, and can work in a variety of useful ways, including identifying the natural language of the document, its media type, its expiration date, and the date on which it was last modified. Some helpful examples of HTTP-EQUIV use, that prove especially relevant for search engines, include:

- **charset** can identify the character set the browser should use, which can prove especially helpful when working with international sites, since many of the search engines now cater to global audiences:

```
<META HTTP-EQUIV="charset" CONTENT="iso-8859-1">
```

- **refresh** helps if you rename your site and want to ensure that search engines can find your new site. This tag will open a new page after one second elapses, and is also known as a redirect:

```
<META HTTP-EQUIV="refresh" CONTENT="1";URL=
"http://www.topsitesforyou.com>
```

NAME

Sometimes, a blurry line exists between the **NAME** and **HTTP-EQUIV** attributes, but enables the user greater freedom in assigning attributes to the tag. Frequently used **NAME** tags, which may be recognized by search engines and can help with the promotion of your site, include:

- **keywords** identify the keywords you think are appropriate, separated by commas. Remember to include synonyms, slang, etc.:

```
<META NAME="keywords" CONTENT="books, jobs,
buy, sale, yahoo, amazon.com, career, free, search
engines, salary, business, money, internet, Web">
```

- **author** identifies the page author's name and possibly his/her email address:

```
<META NAME="author" CONTENT="Kristina M. Ackley
kackley@topsitesforyou.com">
```

- **description** provides a brief description of your page and, depending on the search engine, may be displayed along with the title of your page. The description should be concise and to the point, especially since many search engines display a maximum of 250 characters for a description:

<META NAME="description" CONTENT="Turn the Internet into a private career counselor. Focus on your job search via key Web sites on skills assessment, resumes and cover letters, networking, interviewing, and salary negotiation. Also research companies and industries, learn to use search engines, observe netiquette, and more.">

◆ **robots** tells spiders/Web crawlers which pages they can or cannot index and which links should or should not be followed:

<META NAME="robots" CONTENT="all, index, follow"></head>

Search Engine Submissions

Search engine submission presents some complicated, challenging steps for any Web owner. As discussed earlier, in order to ensure Internet users can locate your site, you must submit information to the top search engines. You can submit your site through automatic tools, which include online tools, such as Submit-It (www.submit-it.com), or purchase software like SubmitWolf (www.submitwolf.com). These tools usually require you to fill out a standard form, indicating information included in your meta tags, such as keywords, description, etc. Then they submit the same form to a variety of popular and less known sites, enabling you to quickly and easily broadcast information about your site.

Many people advocate against using these tools, arguing that you can improve your rankings on search engines if you manually fill out each search engine's submission forms. If you choose to follow this path, you most likely will provide the same information as you would provide using the automated tools; you'll just encounter a stronger probability that the information is applied to the correct fields and that non-standard fields also receive your attention. Regardless of which path you follow, first complete a little research about the top search engines. Make sure that you know the top keywords surfers use to locate Web sites. Also, avoid submitting search engine entries more than once per month, as many search engines will "blacklist" your site if you submit entries too often.

Search Engine Watch
www.searchenginewatch.com
Search Engine Watch advises visitors of the trends and tools that help maximize Web sites' search engine potential. Not only does it identify the most popular search engines, it also rates and reviews them, and identifies the submission guidelines for each. Additionally, you can subscribe to a free monthly newsletter that highlights information you won't want to miss.

Global Spider
www.globalspider.net
Although heavy on advertising for its search engine submission product, Global Spider provides a variety of helpful tips and hints to help you maximize your search engine results. You can subscribe to the site's free bi-monthly newsletter or open the page www.globalspider.net/tips.htm for some great advice.

Primer on Search Engine Use In order to effectively submit information to search engines, you should understand how users utilize the top engines to meet their needs. Search engines provide valuable help to everyone, from the newest novice to the most experienced Internet user. They search the Internet for results in a variety of different ways, from submitting questions to specially formatted keywords and combinations.

Search engines can be a bit tricky for the uninitiated, but with a little practice, users can find anything online. The first step is to choose which words you want to look up. For this search, we'll look for help with CSS style sheets. The second step is to determine the search engine's protocol for searches. Some search engines require users to use "and," "or," and "not" to identify how to search. Others use + or - to distinguish which words to include in your search. Many search engines accept both formats, and most enable you to search for phrases by enclosing the text in quotes. All of them should have a clearly identifiable area for more help on the best way to search on that engine. Depending on the search engine, you can play around with a variety of combinations for your search, such as:

"CSS style sheets"

+CSS +style +sheet

CSS and style and sheet

+CSS +"style sheet"

Note: Certain words, such as "Web" and "Internet" appear so prevalently on the Internet that the search engine may refuse to search for them, or will notify you that an exorbitant number of results exists for that particular word. As this book was being written, searching for "Web" produced over 148 million results on AltaVista and 114 million on Google. Searching for +Web +"css style sheet" on these sites, however, produced 7,173 on AltaVista and 1,970 on Google. Narrowing your search definitely produces more manageable results.

Online Newsletters

Let's face it...people seek information on the Web that provides value to them, but frequently meet with so much completely useless information, they lose faith in the Internet as a reliable resource. Online newsletters can help them filter out the rubbish and establish more control over their Web experience, by providing them with information specifically targeted to their needs. They enable the Web site owners to "push" interesting, useful information to their audience on a regular basis and, in the process, remind them to visit specific Web sites. They also provide a tool through which you can generate revenue, by listing advertising for other sites; or, conversely, you can advertise in someone else's newsletter.

Your Own Online Newsletter

An effectively produced online newsletter might include some marketing information, such as special sales or discounts, but first and foremost should seek to provide users with information relevant to their needs. Research your audience to determine their interests and the "hot topics" that really cause them to take notice. By incorporating links for their Web sites into the subject material, newsletters serve to remind users of the valuable resources available on your Web site, and drive them back to your site. For example, John Kremer, a well-known book marketer, offers an informative email newsletter: "John Kremer's Book Marketing Tip of the Week" (www.bookmarket.com). His target audience—publishers, writers, and others who market books—receives a variety of useful hints and tips, including general news about the publishing industry, media contact updates, reader feedback, and more. Kremer cleverly entwines this information with advertising for his book, Web site, and seminars, thereby capturing readers' interest by presenting pertinent information and telling them how to increase their knowledge in those areas.

Make sure that your newsletter achieves its described purpose. If you promote your newsletter as an invaluable source of information about your industry, but it only covers promotional material about your products, it will most likely result in dissatisfied customers and canceled subscriptions. Additionally, to increase interest and provide a preview for those hesitant about signing up for more email, archive your newsletters in an easily accessed area on your Web site. Usually, you can create a general page that offers users the opportunity to browse old newsletters and subscribe for future editions.

Creation and Distribution In order to create an online newsletter, you will need to follow a few simple steps. First, determine how you will create and distribute your newsletter. If you don't mind a bit of extra work, you can create a list of email addresses, to which you will manually add subscriptions as they arrive, and use the bcc (blind carbon copy) function in your email program every time you want to send out an email. On the other hand, you could work through one of the free online services, like Yahoo Groups (http://groups.yahoo.com) or Topica (www.topica.com), to more effectively manage your lists. If you can access financial support for your newsletter, you might establish a special system through which subscriptions can be automatically captured and maintained.

After you determine your method of capturing users, simply create the newsletter and send it to the list of subscribers. When you design your newsletter, try to avoid using graphics and backgrounds, as they will take longer to download into your subscribers' mailboxes and, depending on their mail settings, may even be completely refused due to their size. Instead, create a simple, text newsletter that can be read by any email program, and make sure it includes links to relevant areas on your site or other sites. Determine the frequency with which you will produce and distribute the newsletter, and ensure that your resources will enable you to stick to that timeline. Most importantly, make sure you always provide a quick, easy, effective option through which subscribers can cancel their newsletter delivery.

Newsletter Promotion Next, determine how you will attract subscriptions for your newsletter. When you create newsletters like these, you can promote them through a variety of methods. First and most obviously, offer them on your Web site, in conjunction with the material they will support. If you sell products, you can provide links to your newsletter subscription page on the bottom of those product pages, or on the sales confirmation page. On the other hand, if you produce a newsletter highlighting interesting facts relevant to the material on your genealogy site, you might create a more prominent newsletter link so people can access that information from any area on your site.

You can also promote your newsletter outside of your Web site. When relevant, advertise through print materials that feature your Web site. As appropriate, visit online newsgroups and listservs and offer it as a resource to interested individuals (or, if you gain permission from the owner or moderator of the listserv or newsgroup, see if you can post the newsletter to the full audience). You can also submit your newsletter to search engines, or use tools like Liszt (www.liszt.com) and Topica as additional promotion.

Riding the Coattails of Others' Newsletters

If you cannot find the time or resources to create and maintain your own online newsletter, you can easily use external newsletters as an advertising tool for your Web site. Usually, one of the most inexpensive ways to work with external newsletters involves your submission of articles relevant to the newsletters, with a link at the bottom of the newsletter identifying its origin. Frequently, since the newsletter owners want to build up their information with interesting, outside content, you may be able to utilize this great resource for free. You can also sponsor other newsletters, helping them defray their costs and receiving a nice link to your Web site somewhere within the newsletter. Whichever method you choose, you determine whether the newsletter offers an appropriate means for reaching your target audience, and ensure that it follows some of the guidelines in the preceding section, "Your Own Online Newsletter." Also, to determine ROI (return on investment), make sure you establish measurements that will enable you to determine the number of users who read the newsletter and the number of hits that come from that newsletter.

Other Marketing Tools and Tricks

Be familiar with your online community…join a listserv or newsgroup, such as those described in Chapter 1, Getting Started, to become a part of the community and establish yourself and your company as an expert when specific questions are asked. Avoid spamming the list with advertisements, as the community will quickly expunge you from the list. Additionally, by having a signature file at the end of your email messages (5-6 identification lines that include your name, contact information, and Web URL), you subtly promote your Web site with every message you compose.

Also, survey your competition to determine their advertisement methods. Do they pay for higher search engine placements? Take a close look at your financial resources and the cost of paid search engine placements. Perhaps you can use the same strategy. Perhaps your peers consistently win awards for their Web site. You'll quickly realize that you can only win awards if you submit entry forms for them, so identify the top Internet award sites and start improving recognition and respect for your Web site. Also, browse through the Web, reviewing the numerous places on which your peers' sites are advertised. In some cases, they may pay for banner or link placement on another Web site. On the other hand, depending on the focus of the site, they may receive free placement in exchange for information or placement on their site. Your Web site's recognition may be limited only by your financial situation and willingness to link to other sites. As you finish this book, take a moment to look at these final sites for more information on successful Web site promotion.

Internet Marketing Center
www.marketingtips.com

Although this site features an advertising-heavy interface, you can gain some very useful advice about online strategies, if you know where to look. The hot articles and free "Killer Tips & Tricks" lessons walk you through some very effective methods for increasing notice of your site. You can also subscribe to the free Internet Marketing Tips Newsletter.

101 free tips.com
www.101freetips.com

This bright site provides just what it states—101 free tips to help you in your marketing efforts. Beyond search engine optimization, learn how to use pop-up windows and email marketing to increase your Web site's reach and impact. You can also subscribe to MarketGold TIP SHEET, a weekly online newsletter.

Appendix of Internet Sites

The following sites were featured in the chapters of this book and should be a useful quick-reference resource during your Web development. They are categorized according to focus and alphabetized within their categories. Hundreds of other sites exist on the Internet to provide you with a wealth of information for online Web design.

Accessibility

Bobby . . . . .www.cast.org/bobby

The Productivity Works of Trenton . . . . .www.prodworks.com

WebABLE . . . . .www.webable.com

Web Page Accessibility Self-Evaluation Test 2.0 . . . . .www.psc-cfp.gc.ca

Animation Software and Resources

Alchemy Mindworks, Inc. . . . . www.mindworkshop.com

Animation World Network . . . . www.awn.com

Lycos Computers . . . . .www.lycos.com

Audio

Adobe . . . . www.adobe.com

Archive Service of the School of Computing, Information Systems and Mathematics,

South Bank University, London, U.K . . . . .http://archive.museophile.sbu.ac.uk/audio

Freebie Music . . . . www.freebiemusic.com

Macromedia . . . . www.macromedia.com

Microsoft . . . . www.microsoft.com

Real.com . . . . www.realaudio.com

Shockwave.com . . . . www.Shockwave.com

Web Reference.Com . . . . www.webreference.com

Yahoo's Audio Index . . . . www.yahoo.com/Computers_and_Internet/Multimedia/Audio

Browsers

Internet Explorer . . . . .www.microsoft.com/windows/ie

Lynx . . . . http://lynx.browser.org

Netscape . . . . http://home.netscape.com

Opera . . . . www.opera.com

Browser Testing

Browser News . . . . http://browserwatch.internet.com

Web Monkey—Sizing up the Browsers . . . . www.webmonkey.com

Clipart

AAA Clipart.com . . . . www.aaaclipart.com

All Free Original Clipart . . . . www.free-graphics.com

Animation Factory . . . . www.animfactory.com

Color

Firelily Designs . . . . www.firelily.com/opinions/color.html

Pantone . . . . www.pantone.com

Color Charts

Brobst Systems www.brobstsystems.com/colors1.htm

Color Matters . . . . .www.colormatters.com

Symbolism   http://webdesign.about.com/compute/webdesign/library/weekly/aa070400b.htm

HEX Color Chart . . . . www.interlog.com

Super Color Chart . . . . www.zspc.com/color

Content Management

BroadVision . . . . .www.broadvision.com

TeamSite . . . . .www.interwoven.com

Vignette . . . . .www.vignette.com

Copyright Laws

The Copyright Website . . . . .www.benedict.com

Library of Congress . . . . .www.loc.gov/copyright

Design—Good and Bad Examples

The Web 100 . . . . www.web100.com

Web Sites that Suck . . . . www.websitesthatsuck.com

World Best Websites . . . . www.worldbestwebsites.com

Domain Registration

Great Domains . . . . .www.greatdomains.com

The Internet Corporation for Assigned Names and Numbers . . . . .www.icann.com

Network Solutions . . . . .www.networksolutions.com

Fireworks

EhandsOn . . . . .www.ehandson.com

Macromedia . . . . www.macromedia.com/software/fireworks

Flash

The Flash Academy . . . . www.enetserve.com/tutorials

Flash Kit . . . . www.flashkit.com

Flash 99% Bad www.useit.com/alertbox/20001029.html

Frames

Framing the Web . . . . http://webreference.com/dev/frames

Johnny's HTML Headquarters . . . . http://webhelp.org/frames.html

Sharky's Netscape Frames Tutorial . . . . www.sharkysoft.com/tutorials/frames

Free Images

EyeWire . . . . www.eyewire.com

PhotoDisc . . . . www.photodisc.com

Genealogy

Genealogy.Com . . . . www.genealogy.com

Cyndi's List . . . . www.cyndislist.com

MyFamily.Com . . . . www.MyFamily.com

Guest Books

Bravenet Web Services . . . . www.bravenet.com

Pathfinder . . . . http://guestbooks.pathfinder.gr

Yahoo's Guest Book Index . . . . http://dir.yahoo.com/Computers_and_Internet/Internet/
World_Wide_Web/Programming/Guestbooks

Hit Counters

Cranfield University . . . . www.cranfield.ac.uk/docs/stats

The Counter.Com . . . . www.thecounter.com

Live Counter Classic . . . . www.chami.com/counter/classic

Microsoft Fast Counter . . . . http://more.bcentral.com/fastcounter

Hobbies

Tripod . . . . www.tripod.com

Yahoo! GeoCities . . . . www.geocities.com

Yahoo! WebRing . . . . www.webring.org

HTML

The Bare Bones Guide to HTML . . . . http://werbach.com/barebones

Free Site Templates . . . . www.freesitetemplates.com

HTML Goodies . . . . http://htmlgoodies.earthweb.com

HTML Cheatsheet . . . . www.webmonkey.com/reference/html_cheatsheet

Illustrator

Adobe's Illustrator site . . . . www.adobe.com/illustrator

Illustrator Resources . . . . www.illustrator-resources.com

Intellectual Property Law

Bitlaw . . . . www.bitlaw.com

NOLO Self-Help Law . . . . www.nolo.com

Intranet Information

CIO's Intranet/Extranet Research Center . . . . www.cio.com/forums/intranet

Intranet Design Magazine . . . . http://idm.internet.com

Java

Freewarejava.Com . . . . http://freewarejava.com

Java Boutique . . . . www.javaboutique.com

The Source for Java Technology . . . . http://Java.sun.com

JavaScript Resources

JavaScript.Com . . . . www.javascript.com

The JavaScript Source . . . . http://javascript.internet.com

JavaScript Tutorial for the total non-programmer . . . . http://webteacher.com/javascript

Listservs

Liszt . . . . www.liszt.com

Yahoo Groups . . . . http://groups.yahoo.com

Meta Tags

Spider Food's Meta Tag Tutorial . . . . http://spider-food.net/meta-tags.html

Vancouver Dictionary of Meta Tags . . . . http://vancouver-webpages.com/META

Newsgroups

Google Groups . . . . .http://groups.google.com
Liszt's Usenet Newsgroups Directory . . . . www.liszt.com/news

Object-Oriented Programming (OOP)

Don't Fear the OOP http://sepwww.stanford.edu/sep/josman/oop/oop1.htm
The Journal of Object-Oriented Programming . . . . www.joopmag.com

Online Newsletters

"John Kremer's Book Marketing Tip of the Week" . . . . .www.bookmarket.com
Liszt . . . . .www.liszt.com
Topica . . . . www.topica.com
Yahoo Groups . . . . .http://groups.yahoo.com

Paint Shop Pro

Jasc software . . . . www.jasc.com
Lori's Web Graphics . . . . .http://loriweb.pair.com

Patents and Trademarks

Trademark.com . . . . www.trademark.com
U.S. Patent and Trademark Office . . . . www.uspto.gov

Personal Web Sites

About.com's Personal Web Pages Site . . . . www.personalweb.about.com
Online Success for Internet Business . . . . www.webmastercourse.com

Photoshop

Adobe's Photoshop site . . . . www.adobe.com/products/photoshop
Photoshop Paradise . . . . www.desktoppublishing.com/photoshop.html

Plug-Ins

Flash Player . . . . www.macromedia.com

QuickTime . . . . www.apple.com

RealPlayer . . . . .www.real.com

Shockwave . . . . www.macromedia.com

Windows Media Player 7 . . . . www.microsoft.com

Polls and Surveys

Infopoll . . . . www.infopoll.com

WebSurveyor . . . . www.websurveyor.com

Professional Organizations

American Society of Association Executives . . . . www.asaenet.org

The Marketing Resource Center . . . . www.marketingsource.com

Search Engines

AltaVista . . . . .www.altavista.com

AOL Search . . . . http://search.aol.com

Ask Jeeves . . . . www.askjeeves.com

Excite . . . . www.excite.com

Google . . . . www.google.com

Hotbot . . . . www.hotbot.com

Lycos . . . . www.lycos.com

Netscape . . . . www.netscape.com

Northerrn Light . . . . www.northernlight.com

Yahoo . . . . www.yahoo.com

Search Engine Software

Ask Jeeves Business Solutions . . . . http://business.ask.com/software

Google Customer Site Search . . . . .www.google.com/services

HotBot Help Tools . . . . http://hotbot.lycos.com/help/tools

Inktomi Internet Search Solutions . . . . www.inktomi.com/products/search

Search Engine Submissions

Global Spider . . . . www.globalspider.net

Search Engine Watch . . . . www.searchenginewatch.com

Submit-It . . . . .www.submit-it.com

SubmitWolf . . . . .www.submitwolf.com

Sites That Supply Content

Moreover . . . . www.moreover.com

iSyndicate . . . . www.isyndicate.com

Subportal . . . . www.subportal.com

Yellowbrix . . . . www.yellowbrix.com

Software and Hardware Purchases

Computershopper.com . . . . www.computershopper.com

Ebay . . . . www.ebay.com

Spelling, Grammar, and Punctuation

EEI Communications . . . . www.eeicommunications.com

The University of Victoria's Hypertext Writer's Guide . . . . . www.clearcf.uvic.ca/writersguide

Style Guides

AT&T: att.com Style Guide . . . . www.att.com/style

Carnegie Mellon Web Publishing Style Guide . . . . www.cmu.edu/home/styleguide

Text Editors

Allaire: HomeSite . . . . .www.allaire.com/Products/HomeSite

Bare Bones Software, Inc . . . . www.barebones.com

BBEdit Tips and Tricks . . . . http://any.browser.org/bbedit

The GNU Emacs page . . . . www.gnu.org/software/emacs

Marjolein's Help for HomeSite Users . . . . www.hshelp.com

Notepad . . . . www.notepad.org

The Unix Reference Desk . . . . <u>www.geek-girl.com/unix.html</u>

The VI Lovers Home Page . . . . <u>www.thomer.com/thomer/vi/vi.html</u>

Testing Sites and Software

Browser Photo . . . . <u>www.netmechanic.com</u>

HTML Tidy . . . . <u>www.w3.org/People/Raggett/tidy</u>

HTML Toolbox . . . . <u>www.netmechanic.com</u>

Lift Online . . . . <u>www.usable.net</u>

WebKing . . . .<u>www.thewwebking.com</u>

W3C HTML Validation Service . . . . <u>http://validator.w3.org</u>

Testing Usability

SURL Optimal Web Design . . . . <u>http://wsupsy.psy.twsu.edu/optimalweb</u>

Usable Web . . . . <u>http://www.usableweb.com</u>

Usability

Jakob Nielsen's site (Usable Information Technology) . . . . <u>www.useit.com</u>

Usability Professionals Organization . . . . <u>www.upassoc.org</u>

Video

Apple's Final Cut Pro . . . . <u>www.apple.com/finalcutpro</u>

ESPN . . . . <u>http://espn.go.com</u>

Microsoft's NetShow . . . . <u>www.netshow.com</u>

Multimakers . . . . <u>www.multimakers.com/video</u>

Streaming Media World . . . . <u>www.streamingmediaworld.com</u>

Windows Media . . . . <u>www.microsoft.com/windows/windowsmedia</u>

Video on the Web . . . . <u>www.yourvideoontheweb.com</u>

Web Design Awards

WebAwards . . . . <u>www.webaward.org</u>

The Webby Awards . . . . <u>www.webbyawards.com</u>

Web Design Job Hunting

Careerbuilder.com . . . . .www.careerbuilder.com

Headhunter.net . . . . .www.headhunter.net

Monster.com . . . . .www.monster.com

Web Portals

Everdene . . . . www.everdene.com

GreatStart.Com . . . . www.GreatStartPage.com

Web Resumes

Resumania . . . . www.resumania.com

4resumes www.4resumes.com

Web Site Promotion

Internet Marketing Center . . . . www.marketingtips.com

101 free tips.com www.101freetips.com

Web Standards

World Wide Web Consortium www.w3c.org

WYSIWYG Web Page Development Tools

Adobe's GoLive site . . . . www.adobe.com/products/golive

Dreamweaver Depo . . . . www.andrewwooldridge.com/dreamweaver

FrontPage World . . . . www.frontpageworld.com

GoLive Heaven . . . . www.goliveheaven.com

ImageCafe . . . . www.imagecafe.com

Macromedia's Dreamweaver site . . . . www.macromedia.com/software/dreamweaver

Microsoft FrontPage . . . . www.microsoft.com/frontpage

Netscape . . . . http://home.netscape.com

The Netscape Unofficial FAQ . . . . www.ufaq.org

Silly Dog Netscape Browser Archive . . . . http://sillydog.webhanger.com/narchive

Index

The Click and Easy™ Online Resource Centers

Books, videos, software, training materials, articles, and advice for job seekers, employers, HR professionals, schools, and libraries

Visit us online for all your career and travel needs:

www.impactpublications.com
(bookstores and Impact Publications)

----

www.winningthejob.com
(career articles and advice)

----

www.contentforcareers.com
(syndicated career content)

----

www.veteransworld.com
(military transition articles and advice)

----

www.ishoparoundtheworld.com
(unique international travel-shopping center)

----

www.contentfortravel.com
(syndicated travel content)